Scientific and Technical Conference on Innovative Mining Technologies (IMTECH 2020)

IOP Conference Series: Materials Science and Engineering
Volume 1134

Szczyrk, Poland
2 - 6 November 2020

Editors:

Dariusz Prostanski
Bartosz Polnik

ISBN: 978-1-7138-3412-0
ISSN: 1757-8981

Printed from e-media with permission by:

Curran Associates, Inc.
57 Morehouse Lane
Red Hook, NY 12571

Institute of Physics
Dirac House, Temple Back
Bristol BS1 6BE UK

Phone: 44 1 17 929 7481
Fax: 44 1 17 920 0979

techtracking@iop.org

Additional copies of this publication are available from:

Curran Associates, Inc.
57 Morehouse Lane
Red Hook, NY 12571 USA
Phone: 845-758-0400
Fax: 845-758-2633
Email: curran@proceedings.com
Web: www.proceedings.com

TABLE OF CONTENTS

INNOVATIVE MINING TECHNOLOGIES

IMTech 2020 Scientific and Technical Conference

2 – 6 November 2020

Due to severe pandemic restrictions, caused by the SARS – CoV2, an electronic form of the conference was offered. The conference covered a broad spectrum of scientific, research and technical subjects, enabling an exchange of knowledge and professional experience among representatives of academia, research institutes, producers of mining machines and equipment as well as end – users from mines.

The IMTech 2020 Conference was oriented mainly onto innovative mining technologies in the aspect of safety, reliability and efficiency.

During twelve conference sessions forty – four papers, concerning work safety in the mining industry, minerals' extraction systems, state–of–the–art control, monitoring and diagnostics systems of machines and equipment, cyber –security, new solutions of hydraulic systems, new generation, highly efficient haulage assemblies for longwall systems, hoisting machines, recovery of minerals, preparation systems, an improvement of mining personnel's qualifications and an internationalization of the Polish circum–mining sector were presented.

In total about 4000 on-line participants took part in the conference. It is worth mentioning that 830 people followed the session simultaneously translated into English. Foreign participants were from Australia, Chile, India, Romania and Slovenia. This conference publication contains only a few selected articles of innovative character, presenting occupational health and safety issues, innovative concepts of minerals' production, as well as test results and analyses of systems applied in mining machines and equipment.

Editors

Dr. Dariusz Prostański, Prof. at KOMAG

Dr. Bartosz Polnik

Scientific Committee of the conference:

IOP Conf. Series: Materials Science and Engineering 1134 (2021) 011001 doi:10.1088/1757-899X/1134/1/011001

Peer review declaration

All papers published in this volume of *IOP Conference Series: Materials Science and Engineering* have been peer reviewed through processes administered by the Editors. Reviews were conducted by expert referees to the professional and scientific standards expected of a proceedings journal published by IOP Publishing.

- **Type of peer review: Double-blind**
- **Conference submission management system:** e-mail
- **Number of submissions received:** 9
- **Number of submissions sent for review:** 7
- **Number of submissions accepted:** 7
- **Acceptance Rate (Number of Submissions Accepted / Number of Submissions Received X 100):** 78%
- **Average number of reviews per paper:** 2
- **Total number of reviewers involved:** 12
- **Any additional info on review process:**

The form of the review was a "double – blind review process". Each article was addressed to two reviewers.The reviewers were asked to evaluate the article in accordance with the below mentioned criteria:

Does the issue of the article fit into the scope of THE IMTECH (INNOVATIVE MINING TECHNOLOGIES) CONFERENCE?

Does the problem being presented in the article is significant for science?

Is the aim of article clearly specified and realized?

Is the paper an original study (it means, it is not the compilation of former known publications)?

Articles that were approved for publication received a comment:

The article is worth publishing in author's version.

or

The article is worth publishing after corrections suggested by the reviewer.

IOP Conf. Series: Materials Science and Engineering 1134 (2021) 011002 doi:10.1088/1757-899X/1134/1/011002

The reviewers wrote their comments on the review form ("REVIEWER'S COMMENTS") or directly on the received version of the article (without names), and the authors were obliged to make corrections.

- **Contact person for queries: mmalec@komag.eu**

IOP Conf. Series: Materials Science and Engineering 1134 (2021) 012001 doi:10.1088/1757-899X/1134/1/012001

Test of the retractable guidance system installed on the level 960 m in the Leon IV shaft in Rydułtowy Coal Mine, Poland

P Kamiński[1, 4], D Prostański[2] and A Dyczko[3]

[1] Faculty of Mining and Geoengineering, AGH University of Science and Technology, Mickiewicza 30, 30-059 Kraków, Poland
[2] KOMAG Institute of Mining Technology, Pszczyńska 37, 44-101 Gliwice, Poland
[3] Mineral and Energy Economy Research Institute of the Polish Academy of Sciences, J. Wybickiego 7a, 31-261 Kraków, Poland

[4] E-mail: pkamin@agh.edu.pl

Abstract. Rope guidance of the conveyances is a reliable and convenient way of guiding cages and skips in the mine shafts and it is used in numerous underground mines all over the world. However it is not popular in Polish coal mines. Most of the Polish copper and coal mines operate on multiple levels, which requires mine shafts and hoists to support more than one level. In such situation it is necessary to install some kind of chairing system on the levels, i.e. stiff guides in their vicinity. Conveyance's movement through the stiff guides on the level requires its speed reduction, which in turn leads to lowering of the hoist's effectiveness. To avoid such situation, retractable guidance system was introduced on the level 960 in the Leon IV shaft in Rydułtowy Coal Mine, Poland. This paper presents the idea of the retractable guidance system, as well as calculations and measurements carried out to prove that the system is safe and reliable.

1. Introduction

Rope guidance of conveyances is in popular use in numerous mine shafts in different underground mines all over the world. Besides its disadvantages, it does have multiple advantages, associated with its installation and operation, as well as economic benefits. Installation of elastic guidance is much quicker and cheaper than the assembly of stiff guides, similar to its maintenance and potential replacement. Rope guidance allows conveyances to travel with high speed, because of the smooth guidance, without shocks and collisions with the guides [1-3].

However, elastic guidance does have an essential disadvantage, which is a necessity for installation of the stiff guidance on the levels. It is crucial to ensure safety of the operation of the hoist. It is also required by Polish law. Because of this requirement, elastic guidance of the conveyances is rarely used in Polish coal mines and in common use in Polish copper mines. It is caused by multilevel exploitation, which enforces one mine shaft with its hoist to operate on many levels, which in turns requires installation of the stiff guidance on each of these levels [4-6].

The biggest inconvenience of the stiff guidance on the levels is the process of the conveyance's entry onto the guides, because of the conveyance's lateral movements, intensified by conveyance's braking and turbulent air flow in the vicinity of the level. Necessity for conveyance's speed reduction decreases level of effectiveness of the mine shaft and a hoist [7-10].

For purpose of improvement of elastic guidance operation and raise of effectiveness level of the mine shaft and the hoist, retractable guidance system was introduced on the level 960 m in the Leon IV shaft in the Rydułtowy Coal Mine (currently ROW Colliery, Rydułtowy department). Patented retractable guidance system allows the conveyance to travel through the mine level with its full operational speed, if people or materials are transferred to a different mine level or in case of transport to the level 960 m, provides conveyance's stability on that level [11].

The problem with stiff guidance on the mine levels, described above, was neither examined nor solved, because the rope guidance of conveyances is rather rarely used in Polish coal mines, mostly because of its disadvantages listed above. Presented solution is innovative because it utilizes hydraulic cylinder to stabilize the conveyance in very safe manner. Additionally, any major intervention in the conveyance's construction is not required, as it was in similar historic constructions. The construction of the retractable guidance system and the mine level allows installation of swinging bridges, which is important to provide safe and reliable transport of materials to the level 960 m [12-15].

Described solution is a pioneering approach in Polish mining industry. It has helped to raise a level of effectiveness of the Leon IV shaft in the ROW Colliery, Rydułtowy department, where it was used. It might also help to raise safety and effectiveness level in other mine shafts in different Polish underground mines.

2. Design of the retractable guidance system

Retractable guidance system replaced the chairing mechanism of a rope guided cage at level 960 of Leon IV shaft in the Rydułtowy Coal Mine. This system was originally introduced in the project [16] and it was developed to solve the problem of the cage guidance through the level 960. The main reason for development of this system was to reconcile safety with effectiveness. To provide proper level of safety using chairing system it was necessary to reduce speed of the cage from 10 m/s to 0,5 m/s, 100 m before the level 960. The effect of such situation was significant as regards an extension of a single ride time to or from the level 1150 m.

Stiff guides on the level 960 m were replaced with the retractable guidance system, which consist of two pairs of top and two pairs of bottom guides and a supporting structure, comprising of four supporting frames assembled to the shaft lining by additional beams, structurally independent from the existing shaft chairing [17-18].

Retractable guides can be moved between two positions: resting and working ones. The motion is restricted by:

- the element of the supporting frame, named the race – bottom section;
- two articulated links (top and bottom) – top section.

Each of the bottom sections of the retractable guidance system are driven by hydraulic cylinder, assembled to the guide's section and the supporting frame. Top sections of the retractable guidance system are driven by two hydraulic cylinders each.

During the movement of the conveyance between the levels 1150 and zero, all sections of the retractable guidance system are set in resting position (Figure 1).

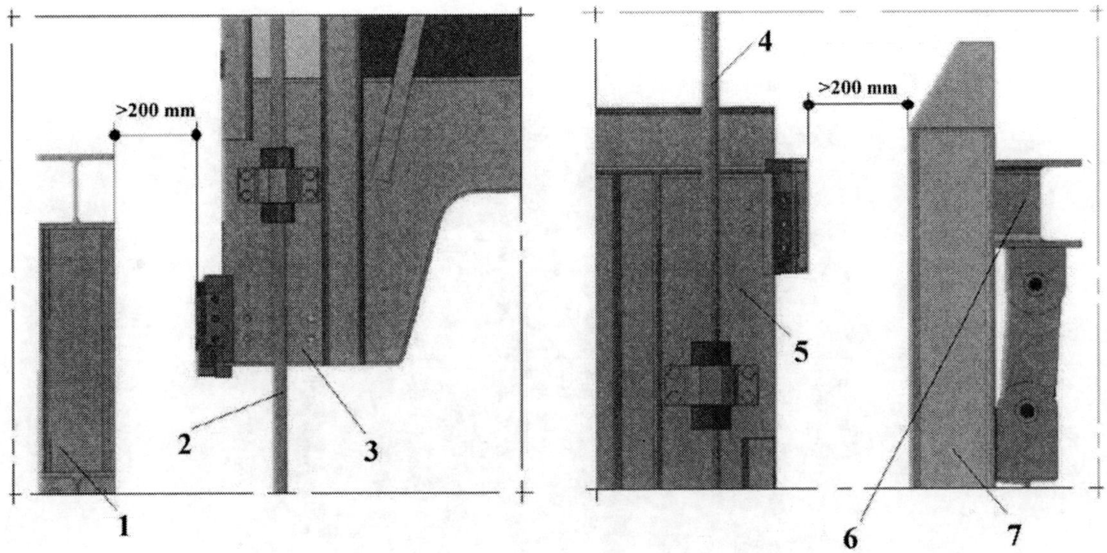

Figure 1. Sections of the retractable guidance system in resting position.
1 – supporting frame of the bottom section, 2, 4 – rope guides, 3 – bottom transom of the cage, 5 – top transom, 6 – supporting frame of the top section, 7 – top section of the retractable guidance system

Retractable guidance system set in resting position allows the conveyance to move through the level 960 with its full speed, which is equal 10 m/s. The system is also set in the resting mode when the cage is moves between levels 960 and 1150 in both directions.

The system is set in the working mode when the cage stops at the level 960. It is done by a banksman on the level 960. Sections of the retractable guidance system in the working position stabilizes the conveyance on the level. Sections of the system are inserted between the elements of the guiding shoes of the cage. Sections of the retractable guidance system in working position are presented in Figure 2 [11, 16].

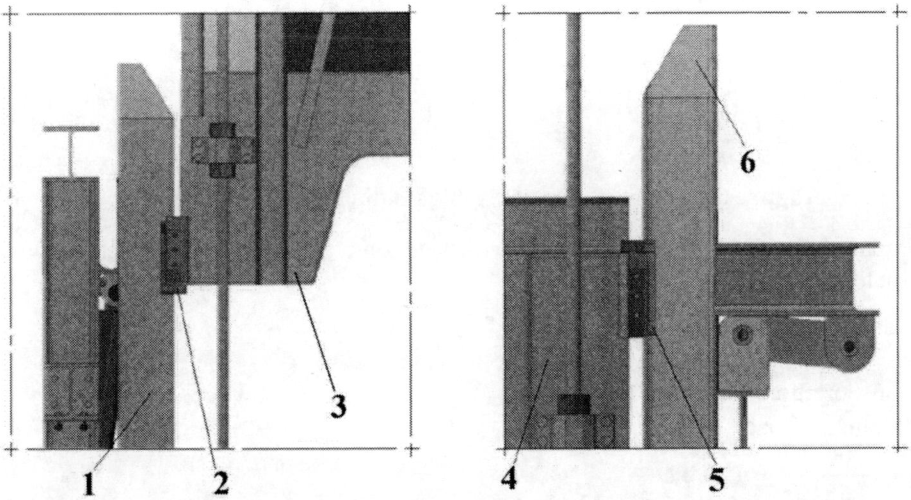

Figure 2. Sections of the retractable guidance system in working position.
1 – bottom section of the retractable guidance system, 2 – guiding shoe, 3 – bottom transom, 4 – top transom, 5 – guiding shoe, 6 – top section of the retractable guidance system

Layout of the retractable guidance system is shown in Figure 3. The sections of the system comprise of 180 x 260 mm box beams made of C260 profiles. Supporting frames, made of HEB 260 are attached to two technological beams with M24 Hex bolts, class 8.8.

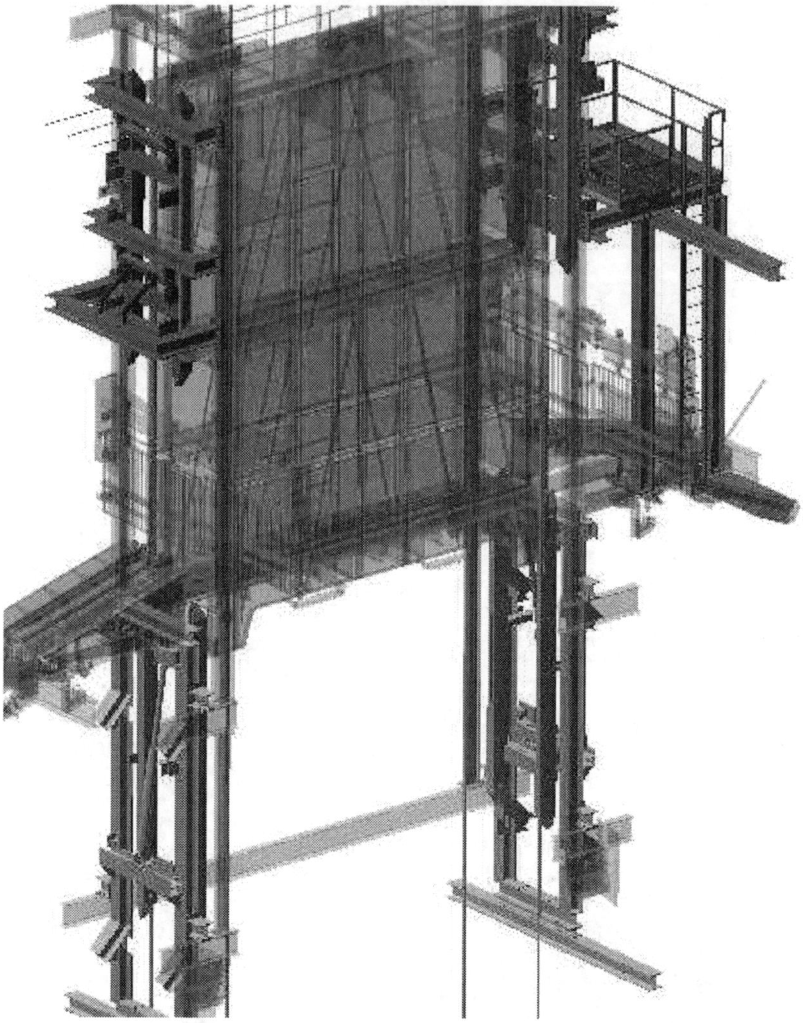

Figure 3. Layout of the adjustable guiding (shown in grey colour).

Static analysis was performed for the constructions. Assumptions, schemes and results are presented below:

STATIC ANALYSIS – BOTTOM SECTION
Input data:

- Transportation unit load: 200 kN
- Continuous load:
  $$q = \frac{200\text{kN}}{1.654\text{m}} = 120.92\frac{\text{kN}}{\text{m}}$$
  Rounded value of the load is equal 121 kN/m
- Factor of safety
  $$\frac{480}{62.87} = 7.63$$
- Force caused by hydraulic cylinder: 139.2 kN

IOP Conf. Series: Materials Science and Engineering 1134 (2021) 012001 doi:10.1088/1757-899X/1134/1/012001

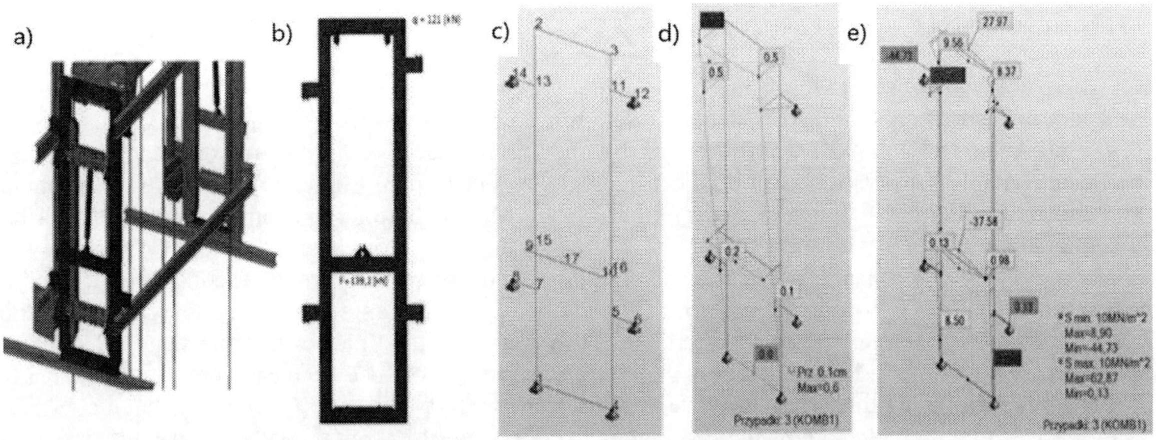

Figure 4. Static analysis of the bottom section:
a) bottom section layout, b) loading scheme of the bottom section, c) model, nodes and constraints, d) loading scheme of the bottom section, e) stress in the bottom section.

STATIC ANALYSIS – TOP SECTION
Input data:

- Transportation unit load: 200 kN
- Continuous load:

$$q = \frac{200\text{kN}}{1.654\text{m}} = 120.92\frac{\text{kN}}{\text{m}}$$

Rounded value of the load is equal 121 kN/m

- Factor of safety

$$\frac{480}{75.09} = 6.392$$

- Force caused by hydraulic cylinder: 139,2 kN

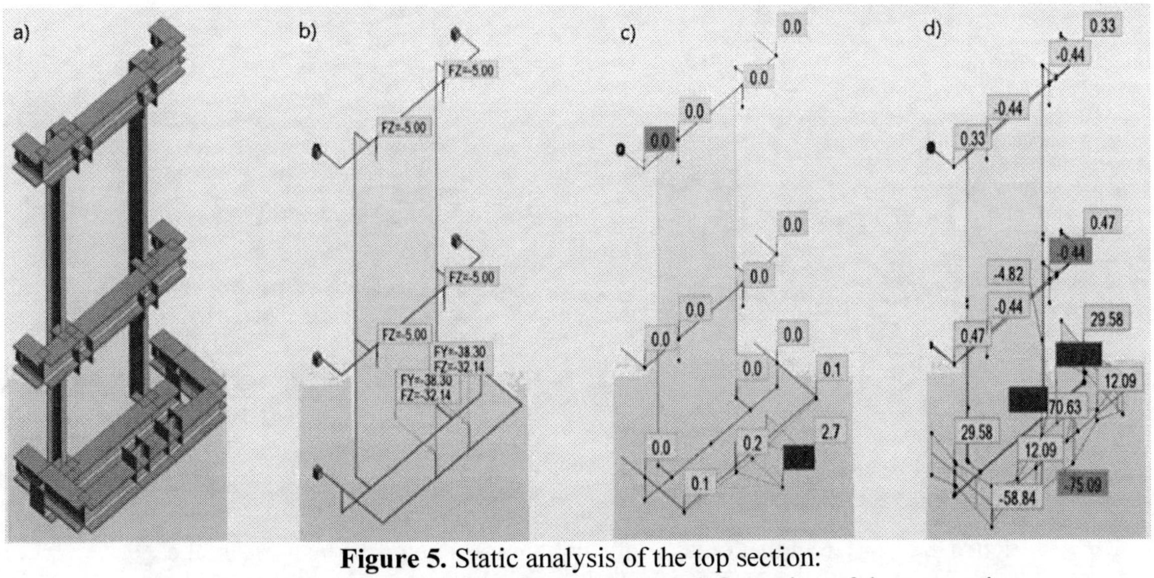

Figure 5. Static analysis of the top section:
a) model of the top section, b) loading scheme, c) deformation of the top section, d) stress in top section elements.

3. Measurements

3.1. Measurements of stiff guides

Presented measurements were conducted to determine peak forces acting on the stiff guides on the level 960 m in the Leon IV shaft, caused by the man-material cage. Results were obtained by measuring value of peak frontal and side forces acting on each of eight stiff guides, four of them cooperating with the top transom and another four with the bottom transom of the cage, during movement of a platform with the total load of 20 tons through a bottom level of the conveyance.

Because of the technical issues, tests were conducted in two stages. First part, consisting of twelve experimental cycles of pushing the platform through the cage, was conducted on the level 960 m of the Leon IV shaft. Another part was performed out of the shaft after the finish of the first stage.

The essence of the first test stage was generation and record of specific measurement signals using specialized instrumentation. The signal represents from mathematical point of view the forces that were to measure. Second stage of the test covered processing of gathered data by computer to obtain measured forces. Method of such processing were presented in several publications, e.g. [19]. Desirable results are obtained using specialized instrumentation, which was previously used in numerous tests in mine shafts. This device allows to achieve uncertainty of measurement of the range of 12 percent.

A special clamping system was designed to attach measuring device to the top and the bottom transom of the man-material cage in the Leon IV shaft. The clamping system allowed the first stage of test to be conducted in the shaft in two phases:

- phase one, conducted to measure forces on the bottom transom of the man-material cage, consists of six experimental cycles of moving the platform through the bottom level of the conveyance,
- phase two, conducted to measure forces on the top transom of the man-material cage, consists of six experimental cycles of moving the platform through the bottom level of the conveyance.

Figures 6 and 7 present diagrams of measured forces. Diagrams presented on the Figure 6 presents forces measured on the bottom transom of the man-material cage, while schemes shown in the Figure 7 presents forces measured on the top transom of the conveyance.

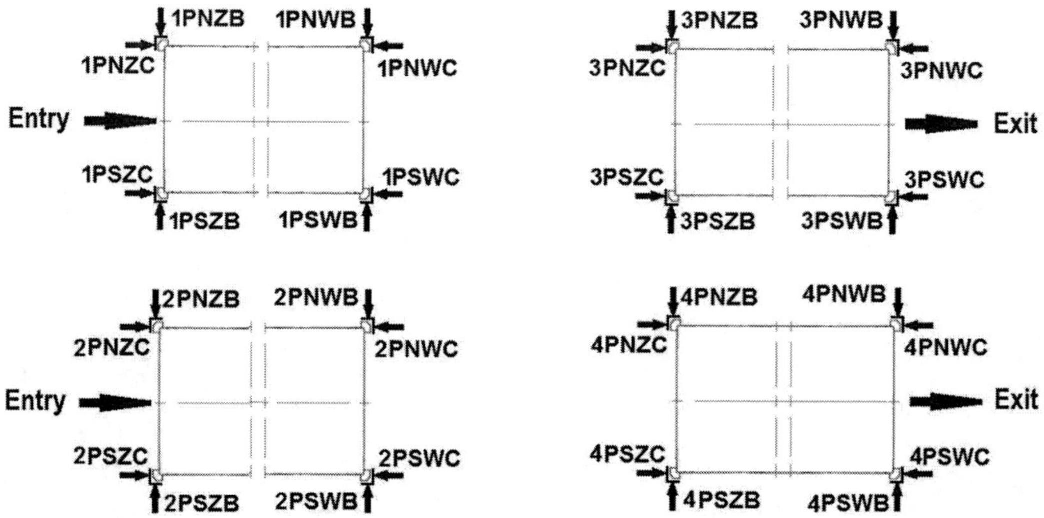

Figure 6. Diagram of the forces on the bottom transom of the man-material cage.

IOP Conf. Series: Materials Science and Engineering 1134 (2021) 012001 doi:10.1088/1757-899X/1134/1/012001

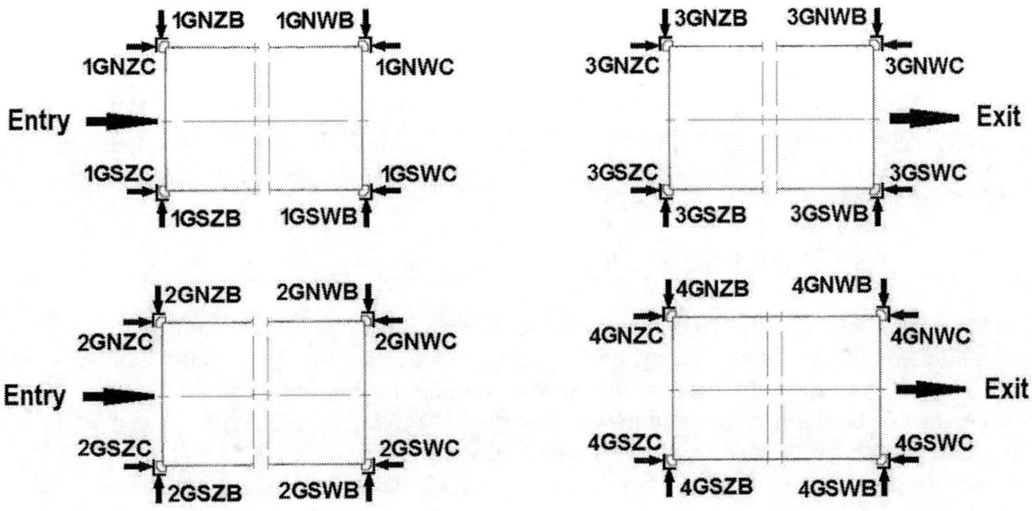

Figure 7. Diagrams of the forces acting on the top transom of the man-material cage.

Markings of measured forces, presented in the figures above, consists of:

- numeric prefix **1**, **2**, **3** or **4**, which match first, second, third or fourth phase of the platform movement through the bottom level of the cage, where:
  - phase **1** comprises of the entrance of the front wheels of the platform onto the cage bottom level,
  - phase **2** comprises of stabilizing the platform in the conveyance,
  - phase **3** comprises of the exit of the front wheels of the platform from the conveyance,
  - phase **4** comprises of the exit of the rear wheels of the platform from the cage.
- letter **P** in each of the marking in the figures 6. and 7. informs that the force was measured on the bottom transom (from Polish *pomost*), while letter **G** – on the top transom (*glowica*).
- letter **N** informs if the force acted on the guide located on the **north**, while letter **S** – on the **south**.
- letter **Z** informs that the force acted on the guide located on the side of the entry of the platform into the cage (*zapychanie*), while letter **W** – on the side of the exit from the conveyance (*wyjazd*).
- letter **C** in each marking corresponds to the frontal forces (*czolowa*) and letter **B** to the side forces (*boczna*).

Tables 1. and 2. present selected peak forces of measured forces, eight for each stiff guide cooperating with the top and bottom transom. Frontal forces are shown in the table 1 and side forces – in the table 2.

Table 1. Peak frontal forces caused by the man-material cage acting on the stiff guidance.

Force marking	Force value, kN							
	1. phase		2. phase		3. phase		4. phase	
	Bottom transom	Top transom	Bottom transom	Top transom	Bottom transom	Top transom	Bottom transom	Top transom
NWC	42.3	28.4	29.7	22.1	48.3	20.0	36.8	44.3
NZC	35.9	21.6	32.5	23.8	35.8	31.4	27.2	25.9
SWC	32.8	29.1	28.5	23.0	36.2	30.3	30.3	48.2
SZC	36.6	36.7	29.5	19.8	31.3	25.4	40.1	24.0

Table 2. Peak side forces caused by the man-material cage acting on the stiff guidance.

Force marking	Force value, kN							
	1. phase		2. phase		3. phase		4. phase	
	Bottom transom	**Top transom**	**Bottom transom**	**Top transom**	**Bottom transom**	**Top transom**	**Bottom transom**	**Top transom**
NWB	14.1	5.5	5.9	5.1	13.1	7.4	20.1	11.5
NZB	15.4	12.5	7.6	**3.3**	8.9	6.1	8.9	8.0
SWB	16.5	7.5	6.4	3.7	16.7	6.2	17.8	18.4
SZB	19.8	9.5	5.8	7.5	17.3	6.7	11.8	7.3

On the basis of tests results presented above, following conclusions were formulated:

1. Values of frontal forces acting on the stiff guidance on the level 960 m and caused by the man-material cage during loading and unloading its bottom level are diverse, because of clearances between guides and the cage as well as random dynamic of:
   a) entry of the platform's front and rear wheels into the cage's level,
   b) exit of the platform's front and rear wheels from the conveyance's level.

   Dynamic's randomness was recognized the main reason of the occurrence of the peak frontal force equal 42.3 kN during loading and 48.3 kN during unloading of the conveyance.
2. Results obtained suggest that peak frontal forces acting on the stiff guides and caused by cage's guide shoe on the top transom should be considered equal to peak forces caused by the guide shoe located near the bottom transom of the cage.
3. Measured peak side forces did not exceed level of 42 percent of peak frontal forces. It should be taken into account that the biggest values of side forces occur in the case of derailing of the platform during its movement through the conveyance, which not happened during the test.

3.2. Measurements of retractable guidance system

Following test aimed in determining peak forces acting on the retractable guidance system on the level 960 m in the Leon IV shaft during loading and unloading of the man-material cage,

Because of the technical conditions, test was parted in two stages. First of them was conducted in the Leon IV shaft and it comprised of generation and record of measuring signals in twelve experimental cycles of moving the platform with the load of 20 tons through the bottom level of the man-material cage. Measuring signals are in a relationship with measured forces and stress. Second stage was carried out after the finish of the first part and it consists of processing of obtained signals. It resulted in values of forces and stress and graphs of their distribution.

Presented tests were conducted similarly to previous measurements, conducted for the case of stiff guidance on the level 960 m and presented in previous section. Retractable guidance system stabilizes the conveyance on the level of its top and bottom transom, it does not provide additional stabilization in the middle of the cage (on the floor of second cage level). Such feature was not considered necessary.

The extension of test was realized by utilization of two additional measuring units, located on the floor of the second cage level, together with the basic measuring units, descripted in the previous section. All of the measuring units are synchronized, thus the uncertainty of measurement is the same as it was in the previous case, so less than 12 percent.

Schedule of the test was similar to the one conducted for stiff guides, presented in previous section. It was divided into two parts"

- stage "Bottom transom", consisting of two series, including three experimental cycles of moving the platform with load of 20 tons through the bottom level of the cage each. During the first part, measuring unit was located on the bottom transom of the conveyance;
- stage "Top transom", consisting of two series, including three experimental cycles of moving the platform with load of 20 tons through the bottom level of the cage each. During the second part, measuring unit was located on the top transom of the conveyance.

In both stages of the experiment, additional measuring units, located on the middle beam of the cage (floor of the second cage level), were used.

4. Comparative analysis of forces measured for stiff guides and for the retractable guidance system on the level 960 m in the Leon IV shaft

For the purpose of the analysis, peak values of compared forces are listed in tables 3 and 4.

Table 3. Value of peak frontal forces acting on the guides on the level 960 m in the Leon IV shaft.

Marking of the force	Peak frontal forces caused by the bottom transom of the man-material cage acting on the guides		Peak frontal forces caused by the top transom of the man-material cage acting on the guides	
	Retractable guides	Stiff guides	Retractable guides	Stiff guides
NWC	38.9 kN	48.3 kN	39.7 kN	44.3 kN
NZC	40.8 kN	35.9 kN	34.9 kN	31.4 kN
SWC	40.7 kN	36.2 kN	40.2 kN	48.2 kN
SZC	38.3 kN	40.1 kN	36.3 kN	36.7 kN

Table 4. Peak side forces acting on the guides on the level 960 m in the Leon IV shaft.

Marking of the force	Peak side forces caused by the bottom transom of the man-material cage acting on the guides		Peak side forces caused by the top transom of the man-material cage acting on the guides	
	Retractable guides	Stiff guides	Retractable guides	Stiff guides
NWB	17.2 kN	20.1 kN	11.0 kN	11.5 kN
NZB	15.0 kN	15.4 kN	12.2 kN	12.5 kN
SWB	17.1 kN	16.7 kN	15.2 kN	18.4 kN
SZB	17.9 kN	19.8 kN	14.1kN	9.5 kN

Table 3 shows that peak frontal forces caused by the bottom transom of the man-material cage and acting on the retractable guides were equal 40.8 kN and on the stiff guides 48.3 kN, respectively forces caused by the top transom for the retractable guides were equal 40.2 kN and for the stiff guides 48.2 kN.

Respectively table 4. shows that peak side forces caused by the bottom transom and acting on the retractable guides were equal 17.9 kN, whereas side forces acting on the stiff guidance were equal 20.1 kN. Top transom caused forces equal 15.2 kN, acting on the retractable guides and 18.4 kN on the stiff guides.

Basing on the presented analysis of the measured forces, one can conclude that values of both peak frontal and side forces caused by bottom and top transom of the man-material cage acting on the construction of the retractable guidance system are significantly lower than the forces acting on the stiff guidance used on the level 960 m in the Leon IV shaft. Such reduction of obtained forces' values may be a result of lower value of the clearance between the retractable guide and the conveyance that between the cage and the stiff guides [20].

5. Comparison of forces measured for the retractable guidance system and valid legislation in terms of safety and Polish Standard PN-G-46227. Mine shafts: Chairing. Requirements.

5.1. Design forces for the retractable guidance system according to Standard

Polish standard PN-G-46227. Mine shafts: Chairing: Requirements [21] does not include definition of the term "retractable guidance system", but because of the analogy between functions of chairing system (defined in the Polish Standard) and retractable guidance system, requirements for chairing system should be considered during measurements and evaluation of the retractable guidance system construction.

IOP Conf. Series: Materials Science and Engineering 1134 (2021) 012001 doi:10.1088/1757-899X/1134/1/012001

Standard [21] consist of following requirements for design forces acting on shaft chairing construction (2.1.1.3):
"Horizontal load acting on the shaft chairing construction during loading a cage with mine cart should be taken equal to peak force acting on a tub creeper's driver, but not less than 10 kN."
"Load-bearing elements of the shaft chairing should be characterized with at least six times high safety factor, without taking wear limit into account."

Cited paragraphs should be considered sufficient for purpose of determination of the design forces of the shaft chairing construction, but only if the system has a typical form of construction. However, construction of the retractable guidance system can be considered shaft chairing system only in terms of its functions, thus determination of design forces according to requirements of the Standard [21] involves additional assumptions for both tub creeper driver's load distribution on different elements of retractable guidance system and selection of design side force acting on these elements.

An attempt to prove the existence of mentioned load distribution was made by comparing peak frontal forces acting on retractable guides and caused by tub creeper – determined as 20 kN [22] with measured peak forces (table 3.). However, this load distribution's existence was not clearly proven. It suggests that peak force caused by tub creeper should be assumed equal to peak design force acting on each of eight retractable guides installed on the level 960 m in the Leon IV shaft.

In terms of side forces acting on the retractable guidance system's elements, it should be assumed, that their construction, different than typical shaft chairing's construction, precludes a priori assumption of compliance of the strength requirements of the side forces, basing on previously presented their compliance in case of the frontal forces. Analysis of side forces revealed that the greatest value of side force acting on the retractable guides occurs in the case of derailment of the transport unit, which had not happened during tests. However, peak side force value in case of the derailment can be estimated basing on the measurement's results. According to this estimation, peak side force value is about 70% higher than measured peak side force and peak frontal force is about 30% lower than measured peak frontal force.

According to facts and assumptions presented above, it was considered reasonable to include two cases during calculation of the forces acting on the retractable guidance system:

- case 1., consisting of design force calculated according to the Standard [21], as the peak force acting on the tub creeper's driver. However, calculations are conducted for case of this force acting on each of the retractable guides,
- case 2., comprising of the transport unit's derailments, with simultaneously acting peak frontal and side forces, equal respectively 70% of peak force on the tub creeper's driver, acting on each of the retractable guides.

5.2. Evaluation of measured forces acting on the retractable guidance system

The base of the evaluation presented below are the safety factor specified in the Standard [21] for the shaft chairing's construction and in the appendix 4. to Regulation [23], in which shaft chairing is considered "auxiliary equipment of the mine shaft".

Required safety factor, presented both in the Standard [21] and Regulation [23] is characterized with equal value, but section 3.15. of the appendix 4. to the Regulation [23] makes this factor a criterion for design stress verification, specifying:

- definition of load-bearing elements' safety factor calculation, as the ratio of the tensile strength R_m of the material and the design stress,
- legitimacy of the Standard's [21] record, which states that wear limit of shaft chairing's elements is not included in their design stresses,
- requirement for corrosion and wear allowance for steel elements of shaft chairing's construction.

Basing on the Standard [21] and requirements of the Regulation [23] obligatory safety factor for load-bearing elements of the shaft chairing was defined as a product of three coefficients:

- n_d – coefficient of frontal dynamic loads acting on the shaft chairing construction.

- n_g – coefficient of limit stress of load-bearing elements of the shaft chairing,
- n_z – coefficient of wear limit of shaft chairing elements.

Coefficient n_d is defined as a ratio of peak frontal force acting on the retractable guides (table 3.) and peak force measured on the tub creeper's driver [22], which is the cause of the force acting on the guides. Calculated value of the n_d coefficient is equal 2.0.

Coefficient n_g was defined because limit stress in the retractable guides is equal to the yield point of the material of which they are made, but verification of stress in the guides should be done basing on its tensile strength, according to the Regulation [23]. Thus, the value of coefficient n_g was accepted at the level of 1.8.

Finally, the coefficient n_z is calculated as a ratio of required safety factor and product of coefficients n_d and n_g. According to this, value of the coefficient n_z is equal 1.67.

Based on values of the coefficients presented above, following note of the forces measured on the retractable guidance system was formulated:

- Peak frontal forces acting on the retractable guidance system, treated as shaft chairing (basing on the existing Regulation [23] and Standard [21]), measured in tests can be considered safe (according to the Regulation [23]), if each of the load-bearing elements of its construction is characterized with safety factor not less than 6 for the peak force caused by the tub creeper's driver acting on each of the eight retractable guides installed on the level 960 m in the Leon IV shaft.
- Peak side forces acting on the retractable guidance system, treated as shaft chairing, measured in tests can be considered safe (according to the Regulation [23]), if each of the load-bearing elements of its construction is characterized with safety factor not less than 6 for the simultaneously acting peak frontal and side forces, equal respectively 70% of peak force on the tub creeper's driver, acting on each of the retractable guides.
- Basing on the calculated safety factors presented above, corrosion and wear allowance (according to the section 3.15.1 of the appendix 4. to the Regulation [23]) can be determined for sections used as the load-bearing elements of the retractable guidance system's construction. To calculate this allowance, following formula is recommended to use:

$$\Delta = 1{,}83 * n_{min}^2 - 5{,}14 * n_{min} - 5{,}18 \qquad (5.1)$$

where:
Δ – limit of wear, as a part of initial section wall thickness, [%],
n_{min} – the smallest safety factor calculated for load-bearing elements of the retractable guidance system's construction.

6. Conclusions

Physical verification of quantities measured in presented tests, which are value of peak frontal forces acting on the stiff and retractable guides revealed that measured forces are satisfactorily reliable image of real forces acting on the elements of the retractable guidance system and the cage in the process of loading and unloading of the cage on the level 960 m in the Leon IV shaft with the transport units of the maximum load possible for the conveyance.

Presented tests demonstrate that:

- peak frontal forces acting on the elements of the retractable guidance system, treated as auxiliary shaft equipment [23] can be considered safe, according to the existing law, if value of design safety factor is not less than six, for the maximum possible force caused by the tub creeper's driver acting on each of the load-bearing elements of the retractable guidance system on the level 960 m in the Leon IV shaft,
- peak side forces acting on the retractable guidance system, treated as shaft chairing, measured in tests can be considered safe (according to the Regulation [23]), if each of the load-bearing elements of its construction is characterized with safety factor not less than 6 for the

simultaneously acting peak frontal and side forces, equal respectively 70% of peak force on the tub creeper's driver, acting on each of the retractable guides.

- Basing on the calculated safety factors presented above, corrosion and wear allowance (according to the section 3.15.1 of the appendix 4. to the Regulation [23]) can be determined for sections used as the load-bearing elements of the retractable guidance system's construction. To calculate this allowance, introduced in the Standard [20] and presented in this work, formula (5.1) is recommended to use.

References

[1] Bura L 1970 *Linowe prowadzenie naczyń wyciągowych w szybach krajowych,* Komunikat GIG nr 483, Wydawnictwo „Śląsk", Katowice 1970

[2] Czaja P and Kamiński P 2017 - *Wybrane zagadnienia techniki i technologii głębienia szybów.* Kraków: Szkoła Eksploatacji Podziemnej

[3] Slonina W and Stuehler W 1980 *Safety problems posed by rope shaft guides, research report,* Commission of the European Communities, Mines Safety and Health Commission, Luxembourg 1980

[4] Olszyna G, Tytko A and Tobys J 2018 Eksploatacja lin prowadniczych i odbojowych, *Napędy i Sterowanie*; 2018 nr 4, s. 110–114

[5] Tobys J and Tytko A 2011 Eksploatacja lin prowadniczych i odbojowych w górniczych wyciągach szybowych, *Transport szybowy 2011 – Monografia*, s. 375-383, ITG KOMAG, Gliwice 2011

[6] Carbogno A and Żołnierz M 2009 Zamocowanie na wieży wyciągowej prowadników linowych za pomocą zacisków klinowych, *Transport szybowy 2009 – Monografia*, s. 151 – 167, ITG-KOAG, Gliwice 2009

[7] Renyuan W, Zhu Z, Chen G, Cao G and Li W 2014 Simulation of the lateral oscillation of rope-guided conveyance based on fluid structure interaction, *Journal of Vibroengineering*, vol. 16, issue 3, 2014, s. 1555-1563

[8] Renyuan W, Zhu Z and Cao G 2015 Computational fluid dynamics modelling of rope-guided conveyances in two typical kinds of shaft layouts, 2015

[9] Chen X 1985 Swing of hoisting conveyance using steel rope guides, *Coal Science and Technology*, s. 23-26, 1985

[10] Taljaard J J and Stephenson J D 2000 State-of-art shaft system as applied to Palabora underground mining project, *The Journal of The South African Institute of Mining and Metallurgy*, 11-12/2000, s. 427-436

[11] Czaja P, Kamiński P, Olszewski J and Bulenda P 2018 Polish experience in shaft deepening and mining shaft hoists extending on the example of the Leon shaft IV in the Rydułtowy mine, *in preceedings of World Mining Congress*, 19–22.06.2018, Astana, Kazakhstan

[12] Bojarski P, Bulenda P, Kamiński P and Nowak J. 2019 Innowacyjny sposób stabilizacji naczyń wyciągowych wyciągu szybowego z prowadzeniem linowym na poziomie pośrednim, *in preceedings of Interntional Mining Forum*, 11-12.04.2019, Katowice, Poland

[13] Kamiński P 2020 Polish experience in shaft deepening and mining shaft hoists elongation, *Mining Techniques-Past, Present and Future*, InTech 2020

[14] Nowak J, Bulenda P, Piszczan Z and Kamiński P 2019 Innowacyjny sposób stabilizacji klatki wielkogabarytowej wyciągu podstawowego szybu Leon IV, *in preceedings of Mechanizacja, automatyzacja i robotyzacja w górnictwie. VI międzynarodowa konferencja,* 12–14.06.2019, Wisła, Poland

[15] Olszewski J, Bulenda P, Chomański B, Jara Ł and Kamiński P 2017 Szyb Leon IV – pogłębianie i wydłużanie górniczych wyciągów szybowych, *Inżynieria Górnicza* nr 1–2, s. 49–52

[16] Projekt techniczny prowadzenia klatki wielkogabarytowej na poz. 960 m (1000 m). Projekt nr Ry-157/218 opracowany przez PBSz S.A. Górnicze Biuro Projektów w 2018 r., Autorzy: B. Chomański, K. Pyrek, P. Kamiński (unpublished)

IOP Conf. Series: Materials Science and Engineering 1134 (2021) 012001 doi:10.1088/1757-899X/1134/1/012001

[17] Olszewski J, Czaja P, Bulenda P and Kamiński P 2018 Pogłębianie oraz wydłużanie górniczych wyciągów szybowych szybu Leon IV w kopalni KWK ROW ruch Rydułtowy, *Przegląd górniczy* vol. 74, nr 8, s. 7 – 16, 2018

[18] Wowra D, Nowak J, Witkowski J, Izydorczyk P and Kamiński P 2017 Wydłużenie Górniczego Wyciągu Szybowego – Szyb Leon *IV in preceedings of IV Polski Kongres Górniczy 2017,* 20–22.11.2017, Kraków, Poland

[19] Płachno M 2005 Nowe metody projektowania i eksploatacyjnej kontroli zbrojenia pionowych szybów górniczych, *Problemy Inżynierii Mechanicznej i Robotyki - monografia*, Wydawnictwo AGH, Kraków, Poland

[20] Polish Standard PN-M-06515:1979. Dźwignice. Ogólne zasady projektowania stalowych ustrojów nośnych (Lifting equipment. Rules of dimensioning of steel structures)

[21] Polish Standard PN-G-46227:2002. Szyby górnicze. Oszybia. Wymagania (Mine shafts. Chairing. Requirements)

[22] Zaprojektowania uzbrojenia szybu Leon IV KWK Rydułtowy-Anna. Poziom 960 m. Urządzenia przyszybowe ze sterowaniem. BSiPG, Katowice 2005 (unpublished)

[23] Rozporządzenie Ministra Energii z dnia 23 listopada 2016 roku w sprawie szczegółowych wymagań dotyczących prowadzenia ruchu podziemnych zakładów górniczych, Dz. U. z 2017 r. nr 1118 (Regulation of the Minister of Energy of November 23, 2016 on detailed requirements for the operation of underground mining plants)

Temporary mechanized gate roof support in the YapiTec mine

J Korski

FAMUR S.A., Armii Krajowej 51, 40-698 Katowice, Poland

E-mail: JKorski@famur.com.pl

Abstract. For more than sixty years in all the world are trials of gateroad temporary mechanized roof supports in mechanized development operation have been conducted unsuccessfully. A case study – temporary mechanized gateroad roof support implemented in one of the entrances/ drifts development in the turkish coal mine Yapi Tec Maden from the construction to 12-month practical experiences in practical use as described in the article. The review of gateroad roof support for different purposes is presented in article as a background or circumstances for the described case study. The YapiTec trial of the company own temporary mechanized gateroad roof support is discussed in comparison with second mechanized gateroad face without temporary mechanized roof support and same mining conditions during comparable time.

1. Introduction

For about 60 years, attempts have been made to implement mechanized roof supports for general use as temporary roof supports in narrow mechanical mining faces, enabling the separation of the mining operations with a narrow face roadheader and making the final gate roof support.

The first such solution was implemented in narrow stopes (stoping faces) in the American state of Pennsylvania, where the final support of the excavation consisted of bolting [1,2]. Later, in the European coal basins of Germany and the UK, numerous attempts were made to apply temporary mechanized face roof supports for working with the roadheaders. The main goal of such actions was to extend the available time of cutting with a roadheader, as the erection of the final roof support (or bolting) in the face required the roadheader cutting operation to be stopped [3,4,5].

It was recognized that the use of a temporary mechanized roof support in the face will allow for the final gate roof support to be made at a certain distance from this face and in a place where there is more space and less mechanical devices in operation. Despite the failures of such attempts, works and real actions have been still undertaken to develop new solutions for mechanized temporary roof supports for the purpose of drilling the gates by means of the roadheaders [6]. It should be noted that the available literature on the subject practically ignores the issue of the use of temporary mechanized gate roof supports in the roadheading [7,8].

The only known and widely applied solution of temporary mechanized gate supports in the roadheading are Bolter-Miner machines produced by several manufacturers, which use different roof supports being an integral part of the roadheader and forming a temporary roof support [9].

General overview of the gate roof supports classified according to their purpose and a case study, i.e. experience with the use of a temporary mechanized gate roof support in one of the currently driven dipheadings providing access to the lignite deposit in the Turkish YapiTec Maden mine near the city of Tuncbilek in the West Anatolia is presented below. This case is very interesting as it is possible to compare the results obtained in two headings being driven in the same mining and geological conditions, but with use of different mechanization systems. On this case study is easier to understand why, after

IOP Conf. Series: Materials Science and Engineering 1134 (2021) 012002 doi:10.1088/1757-899X/1134/1/012002

more than 60 years of trials and experiences of mechanized temporary roof supports anywhere was successfully implemented really useful solution. This article is the trial for explaining reasons of those.

2. Mechanized gate roof supports: types and functions

Nowadays, there are several types of mechanized roof supports in the gateroads (Figure 1), depending on the purpose of their intended use:

- Temporary mechanized roof supports used during the gateroad development/driving as a device enabling the final roof support to be made outside the face area without stopping the cutting operation.
- Mechanized gate roof supports securing the roof at the crossing with the longwall (LW entrance).
- Mechanized gate roof supports securing the gates forward the longwall face in operation [12].

There have been and are still in use various solutions of the gate rood supports intended for various purposes, but temporary mechanized supports in the headings driven with the roadheaders have not been successful.

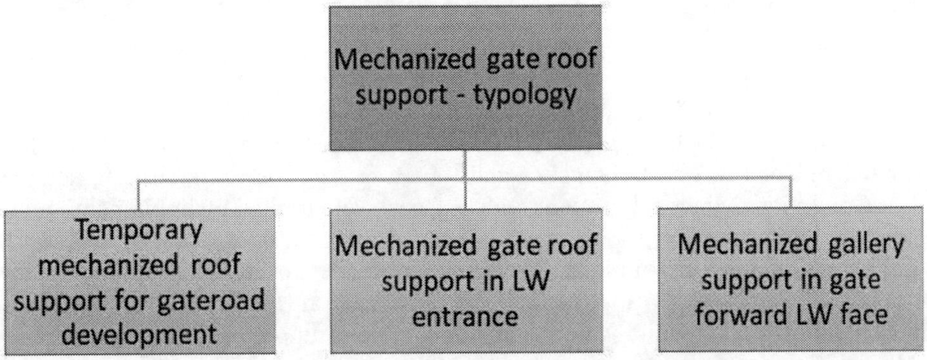

Figure 1. Mechanized gate roof support - typology.

Mechanized gate roof supports, regardless of their intended use, are applied in mining operations with mechanical rock cutting, where there is a potential performing of two operations parallelly: mechanical mining and securing the heading/gate for its expected lifetime.

2.1. Temporary mechanized gate roof supports as protection of the face/heading immediately after exposure

This type of roof supports was already in use in the late 1950s as a way to increase productivity, first in room and pillar systems with the roadheader cutting. The necessity to secure the roof with anchors forced the cutting machine to be stopped during this operation. The solution indicated was the use of a mechanized advancing temporary support that enabled to transfer the roof anchoring operation beyond the face area (behind the temporary roof support area) and to perform it simultaneously with the cutting of the face. The first solution applied was the Dowty Rodjo Shield casing used in the USA as a temporary roof support in the faces of the room & pillar system excavated with a Continuous Miner (Figure 2).

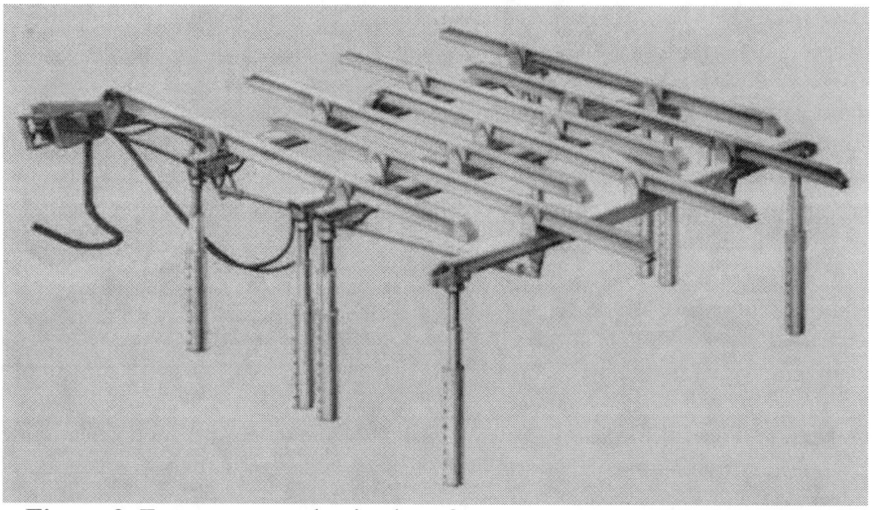

Figure 2. Temporary mechanized roof support - Dowty Rodjo Shield [1].

The implementation of continuous miner machines with a linear cutting head made the Dowty Rodjo Shield roof support lose its technical justification. For this reason, for a better use of the cutting machine - the *continuous miner*, multi-face room & pillar systems were used, where the cutting machine, after exposing the roof to a safe depth, travelled to the adjacent face, and a self-propelled bolting machine drove into the abandoned face to bolt/anchor and thus to secure the roof. Time losses related to the passage of the Continuous Miner between individual faces, were limited by implementing faster driving mode (the so-called tram speed).

Based on these roof support solutions, attempts were made in the British mining industry to use aggregates for drilling roadways with a rectangular cross-section (Figure 3). The mechanized support was a temporary roof support behind which a rectangular yield support was constructed.

Figure 3. Anderson's aggregate for roadway development.
Roadheader with mechanized temporary roof support [9].

In the following years, numerous attempts were made to use temporary mechanized gate supports also for the faces with different types of the final steel support. There were designed and tested different solutions as an independent roof support or as an aggregate – roadheader integrated with temporary roof support (mostly with steel arches as final roof support). As an example Westfalia Lunen designed mechanized temporary roof support in both versions (independent and integrated with roadheader). Voest Alpine Bergbautechnik designed and introduced in Germany AVSA aggregate – milling gallery cutter-loader integrated with temporary roof support. This solution was tested with bolts or steel arches as a final roof support. Any of those solutions anywhere did become widespread [10].

Temporary mechanized roof supports being an integral part of the Bolter Miner machines became widely used [11].

2.2. *Mechanized gate roof support in LW entrance*

Mechanized gate roof support in the longwall entrance became a solution intended to secure the gateroads at the entrance to the longwall. There are many different solutions for such roof supports, one of which is shown in Figure 4.

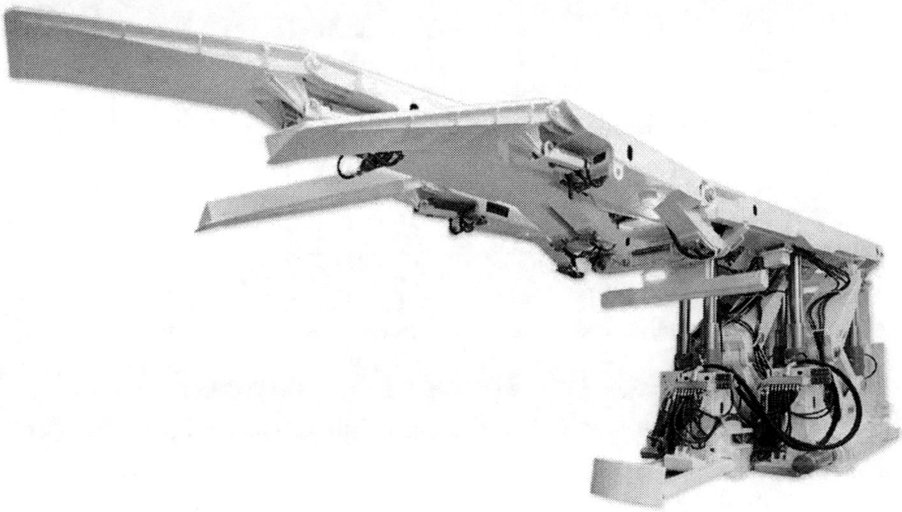

Figure 4. One of typical solutions of gate roof supports from FAMUR S.A.

This type of roof supports did not solve all the problems related to the longwalls with the gates with steel arch supports. An attempt to solve this problem was the mechanized gate roof support with steel arch support at the crossing with the longwall, presented in 2020 [12] (Figure 5).

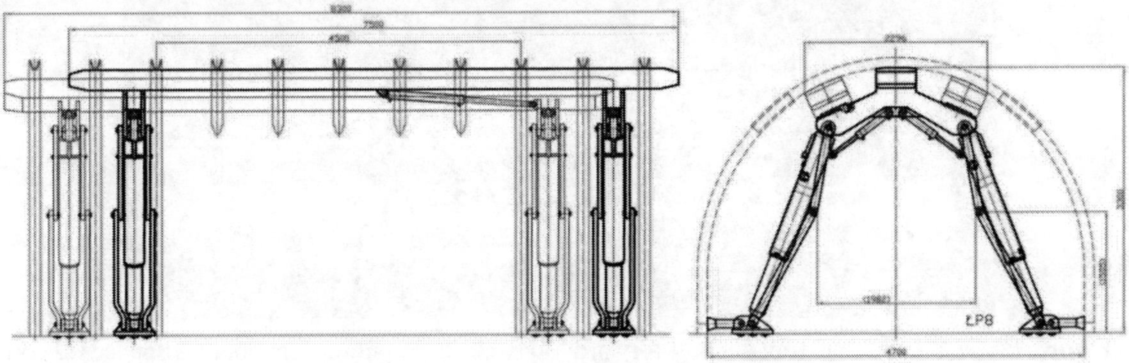

Figure 5. New solution of gate support set [11].

IOP Conf. Series: Materials Science and Engineering 1134 (2021) 012002 doi:10.1088/1757-899X/1134/1/012002

Mechanized gate supports for gateroads with rectangular (or trapezoidal) cross section and with an anchor support are widely used, enabling a full mechanization of operations at the entrance to the longwall.

2.3. Mechanized gallery support in the gate forward to LW face

In Chinese ultra-high (over 6 m high) mechanized longwalls, independent, yield supports were used, advancing in the gateroads ahead of the face front. In these gateroads, various types of anchoring with various anchors/bolts are applied, and for additional protection of such a gate ahead of the face front in the area of the strongest impact of operational pressure, mechanized advancing supports were applied (Figure 6) [12]. Such supports are used in both gates, but they differ depending on the purpose/intended use of the gate.

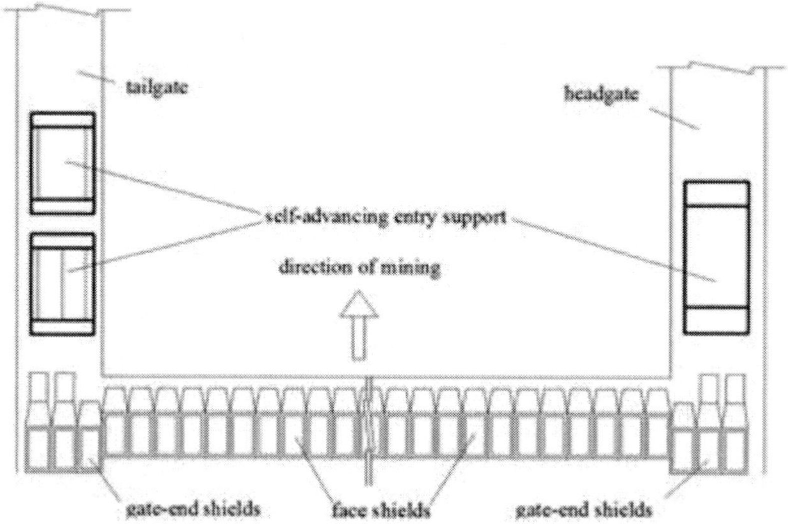

Figure 6. Roof support at longwall face and headgate and tailgate face-end outby T-junctions [13].

The applied supports with a rectangular cross-section and a large free cross-section under the roof support enable a free air flow and an installation/placement of the necessary devices.

Figure 7. Mechanized gate roof support forward longwall face in ultra-thick longwall face (LW) in Chinese coal mine.

IOP Conf. Series: Materials Science and Engineering 1134 (2021) 012002 doi:10.1088/1757-899X/1134/1/012002

Such roof supports enable a safe and reliable operation of very high longwalls while providing high productivity and safety (Figure 7). Ultra-high longwall faces equipment is very expensive

3. Temporary roof support of the dipheading being driven in the YapiTec Maden mine (Turkey)

In 2019, the opening/development of the Turkish YapiTec Maden coal mine deposit in western Anatolia began. The access to the mine will be provided by two dipheadings driven from the surface, which are drilled with roadheaders. The investor has extensive experience in drilling tunnels with reinforced shotcrete lining. This is probably why in one of the dipheadings, a temporary mechanized support for the face was used with the final roof support of reinforced shotcrete (sprayed concrete lining) (Figure 8).

Figure 8. Entrance to drift with temporary mechanized roof support in The YapiTec mine.

The Ukrainian KP-42 (YaMZ) roadheader works in the heading face. Cutting conditions are: soft shales with clay slices or clay for cutting,

3.1. Temporary and final roof support

A combination of several solutions has been selected as the final roof support of the transport dipheading:

- Following the moving mechanized roof support, the roof and the sidewalls of the heading are anchored with steel bolts with the additional welded steel mesh being applied (Figure 9).

Figure 9. Bolter installed on advanced platform behind the roadheader and mechanized temporary roof support.

- Following the anchoring of the sidewalls and of the roof heading, a reinforcement is assembled of welded trusses made of reinforcing rod, in the form of an arched "roof bar" and straight sidewall props (Figure 10).

Figure 10. Prefabricated elements of shotcrete reinforcement.

The main reason for the application of the above-mentioned steel trusses is the planned use of diesel suspended monorail as a means of transport in this mine, and it was found that the anchoring of the

IOP Conf. Series: Materials Science and Engineering 1134 (2021) 012002 doi:10.1088/1757-899X/1134/1/012002

monorail suspension slings in the local conditions (clay and clay-shale) may be insufficient. It was planned to install the truss square sets with spacing of 2 or 3m.

After installing the truss square sets, the roof and sidewalls of the heading are to be covered with shotcrete.

After an exposure of the roof and of the sidewalls, a temporary hydraulic powered support was used as the preliminary protection of the heading.

3.2. Temporary roof support and collaborating devices

The temporary mechanized support applied in the YapiTec Maden mine was developed and produced in the production plant (MarBeyaz Makina), belonging to the same capital group as the mine under construction. The mechanized support is composed of three steel frames with a shape adapted to the intended shape of the heading (straight vertical sidewall parts and an arched roof part), each frame stands on 4 hydraulic props. The frames are connected with each other by means of horizontal hydraulic cylinders (6 cylinders between each pair of frames), these cylinders are intended to move the frames (possible movement in both directions). There is a folding platform/bridge built-in from each side, on the vertical part of the frame, enabling to carry out some maintenance and other operations without an exposure to a contact with the roadheader in operation (Figure 11).

Figure 11. Temporary advancing gallery roof support in entrance drift development I YapiTec Maden.

To avoid a collision with the moving roadheader and its belt conveyor, the advancing working platform has been built behind the mechanized support, with embedded bolting machine and the manipulator for assembling the prefabricated steel reinforcement (Figure 12).

Figure 12. Advancing working platform for bolting and reinforcement assembly with the manipulator.

The main purpose of using this advancing working platform was to enable uninterrupted operation for workers to carry out anchoring and assembling the prefabricated reinforcement in comfortable and possibly stable conditions.

3.3. Temporary mechanized gate roof support and results of its application after 12 months

After 12 months of using the mechanized temporary gate roof support, it is possible to compare the results achieved when it was applied and the results of the dipheading driven with the modernized AM-50 roadheader, but without using a temporary mechanized support in the heading face. AM-50 is smaller and less powerfull than Ukrainian KP-42 in face with mechanized temporary roof support. By September 2020, approx. 300 m of the dipheading had been driven with the use of the mechanized support and approx. 600 m of the other heading. The use of the mechanized support caused (among other reasons) the worse result of the dipheading drivage with this support. The main reasons for the decrease in drilling yield (advancing rate) are specified as follows:

- Low efficiency of the roof and sidewall anchoring system resulting from the type of rocks surrounding the heading (clays, clay shales), resulting in the need to stop cutting operations until anchoring/bolting is completed,
- The movement of the mechanized roof support and the inevitable "pumping" of the roof with the support caused the process of rock stratification to begin and, consequently, there were difficulties with drilling holes for anchoring,
- The cutting capability (advancement) of the roadheader is much greater than the possibility of performing all forms of securing the heading.

4. Conclusion

A low efficiency (advancing rate) of the dipheading drivage with the roadheader and the temporary gate support in the YapiTec Maden mine once again points to erroneous assumptions in the planning of the drilling process with the use of such a mechanization system. In the author's opinion, the application of the temporary gate support in drilling of headings with the roadheader and the use of the final yield support (e.g. arch) is burdened with a logical error (fallacy).

As in previous mechanized, temporary roof support solutions main reasons of problems in Yapi Tec mine are:

- Temporary roof support advancing is "pumping" and fracturing roof strata. As a result roof bolting is difficult and delayed.
- Adopted from tunelling temporary mechanized roof support needs space for assembling and disassembling – in tunnel development drilling is starting from surface to surface. In underground mining at least one of gallery (incline, drift) is underground and it is problem how to exchange temporary roof support to final in the end of gallery.
- There is problem with inclination or direction change with mechanized temporary roof support – it is impossible to change direction or inclinations immediately.
- It's impossible to prepare side entrances from the face in time of drilling gallery with Roadheader and mechanized temporary roof support.

Generally, they are main reasons explaining why mechanized temporary roof support are dead end in gateroad development.

References

[1] https://www.dowtyheritage.org.uk/content/dowty-group (accessed 20-03-20202)

[2] Lesiecki W and Regulski 1957 W *Urabianie złóż. Część 3. Urabianie kombajnami*. Seria: Górnictwo Tom 5. Wydawnictwo „Śląsk", Katowice

[3] Kozdrój M 1972 *Organizacja i podstawy automatyzacji zarządzania w kopalniach węgla kamiennego*. Wydawnictwo „Śląsk", Katowice

[4] Zając E 1994 *Organizacja produkcji w kopalni w kopalni węgla kamiennego*. Śląskie Wydawnictwo Techniczne, Katowice

[5] Yegorov PV., Bober Ye.A., Kuznetsov YU.N., Mikheyev O.V. and Krasil'nikov B.V. 2016 *Podzemnaya razrabotka plastovykh mestorozhdeniy*. "Gornaya Kniga" Izdatel'stvo moskovskogo gosudarstvennogo gornogo universiteta, Moskva

[6] Przybyła H and Chmiela A 2002 *Technika i organizacja w robotach przygotowawczych.* Wydawnictwo Politechniki Śląskiej, Gliwice

[7] Krauze K, Bołoz Ł and Wydro T 2013 Badania stanowiskowe tymczasowej zmechanizowanej obudowy chodnikowej TZOCH-ŁP9 (Testing stand examinations of mechanized roadway temporary support TZOCH-ŁP9), *Problemy bezpieczeństwa w budowie i eksploatacji maszyn i urządzeń górnictwa podziemnego*: *monografia*. Centrum Badań i Dozoru Górnictwa Podziemnego Sp. z o.o. — Lędziny CBiDGP Sp. z o.o. pp.121–132

[8] Peng S S 2020 Longwall Mining 3rd Edition. CRC Press/Balkema, Leiden.

[9] Klich A (ed.) 1999 *Maszyny i urządzenia dla inżynierii budownictwa podziemnego. Wyrobiska korytarzowe i szybowe w górnictwie*. „Śląsk" Sp. z o.o. Wydawnictwo Naukowe, Katowice

[10] Korski J 2020 *Tymczasowe obudowy zmechanizowane w drążeniu wyrobisk korytarzowych.* W monografii: Praca zbiorowa: INNOWACYJNE TECHNIKI I TECHNOLOGIE DLA GÓRNICTWA BEZPIECZEŃSTWO – EFEKTYWNOŚĆ – NIEZAWODNOŚĆ KOMTECH 2020. ITG KOMAG, Gliwice, Poland.

[11] Korski J 2018 *Zastosowanie kombajnów typu Bolter Miner w warunkach polskich kopalń - uwarunkowania i ograniczenia* (Bolter Miners used in the Polish mines), Conference proceeding KOMAG KOMTECH, pp. 5-19, http://komag.eu/images/maszynygornicze1/KOMTECH/KOMTECH_2018_open.pdf

[12] Szyguła M and Mazurek K 2020 Mechanization of reinforcing gate supports in the zone of longwall inlet. *Mining Machines* 1/2020 pp.35-45

[13] Peng S S, Li H, Zhou Y and Cheng J 2013 Ultra-thick seam longwall in China https://www.coalage.com/features/ultra-thick-seam-longwall-mining-in-china/

IOP Conf. Series: Materials Science and Engineering 1134 (2021) 012003 doi:10.1088/1757-899X/1134/1/012003

Analysis of the impact of mining and geological conditions in longwalls on the introduction of pressure monitoring for powered roof supports

K Trzop[1], D Szurgacz[2,5], S Zhironkin[3] and A Kasprusz[4]

[1] KWK Ruda Ruch Bielszowice ul. Halembska 160, 41-711 Ruda Śląska, Poland
[2] Polska Grupa Górnicza S.A., ul. Powstańców 30, 40-039 Katowice, Poland
[3] National Research Tomsk Polytechnic University, 28 Vesennya st., 634034 Tomsk, Russia
[4] Huta Łabędy S.A. ul. Anny Jagellonki 45, 44-109 Gliwice, Poland

[5] E-mail: dawidszurgacz@vp.pl

Abstract. Polska Grupa Górnicza has analysed the longwalls in its mines where powered roof supports should be subject to pressure monitoring. The main goal was to increase the safety of people working in the longwall. The scope also included the economic improvement of active longwalls in terms of stoppages caused by roof fall or possible tremors. The analysis, carried out in this regard, allowed strategic longwalls to be selected from the individual mines of Polska Grupa Górnicza. The paper describes mining and geological conditions in the longwalls, where monitoring is carried out, with particular attention paid to potential tremors. The paper discusses the scope and use of powered roof supports in longwalls. The research team selected the pressure monitoring system as the most important parameter of work of the powered roof support. The paper also presents the results of an analysis carried out on the implementation of the system. A survey was conducted among the employees regarding the relevance of the monitoring system. The analyses results were used to recommend the directions of further actions related to a development of the system for monitoring the supporting parameters of the powered roof support systems.

1. Introduction

In Poland, hard coal is the basic raw material used to generate electricity and heat, thus guaranteeing energy security. Despite numerous changes in the structure of the functioning mining plants, restructuring of the mining industry, automation of exploitation processes, underground mining is an area of economy ranked among the most dangerous in terms of accidents and occurrence of natural hazards [1]. An accumulation of negative factors, such as methane, rock bursts, ejecting rocks, and so on, calls for actions to improve working conditions and comfort [2]. Therefore, it is very important to invest in innovative machinery and equipment to ensure safety, comfort and operating efficiency [3,4,5,6].

Polish hard coal mining is characterised by an occurrence of several natural hazards that necessitate taking preventive measures to improve occupational safety. The application of the proposed solutions largely influences the mining method, efficiency and mining costs [7]. It is accepted that the safety factor is primary and forms several challenges for companies competing in the mining machinery and equipment market. All the activities, especially an implementation of new technical solutions for mining already at the design stage, force companies to take into account geological conditions and natural hazards occurring in mining [8,9].

The article discusses the operation of the pressure monitoring system of a powered roof support unit in selected longwalls of the Polska Grupa Górnicza. In addition, it proposes measures that can be taken to improve occupational safety in mining and, based on surveys, discusses the reaction of employees to the application of innovative solutions in mining using a system for monitoring the units of powered roof supports. The paper aims to analyse the pressure monitoring system against natural hazards and the way the employees perceive it in practice.

2. Characteristics of natural hazards for selected longwalls in Polska Grupa Górnicza covered by pressure monitoring

In the Polish coal mining, practically all the natural hazards typical of underground mining occur, especially the so-called catastrophic hazards. Natural hazards generate several incidents and accidents. The degree of natural hazards present can lead to a reduction in mining efficiency [10]. This phenomenon may even result in the shutdown of a mining plant. Natural hazards are an integral part of the mining operation, but modern machinery, equipment and technologies for mining in difficult conditions make it possible to avoid them [11]. The characteristics of natural hazards in coal mining are presented below [12].

Methane is an odourless, colourless, flammable mine gas and, when combined with air, forms an explosive mixture [13]. The presence of methane in the mine air can displace oxygen, creating an unbreathable atmosphere and an explosion. At concentrations up to 5.0%, methane burns off steadily in contact with a thermal source, in the range 5.0-15% an explosion may occur, and above 15% the mixture is flammable. The strongest explosion occurs at the 9% methane concentration.

Rock bursts are phenomena during which the structure of rocks surrounding workings is violently destroyed as a result of exceeding their strength. It manifests itself in a dynamic movement of rock rubble into the excavation followed by a loud noise with gusts of air and a tremor.

A coal dust explosion is classified as Class A and B. Several factors must be fulfilled for a coal dust/air mixture to explode:

– the coal dust must contain more than 10% volatile matter calculated on the anhydrous and ash-free basis,
– must be fine, its grains not passing through a sieve with a mesh size of 1.0×1.0 mm,
– coal dust must be mixed with air within a range of 50 to 1 000 g/m^3.

Water hazards are divided into three degrees and are influenced by the degree of hydrogeological conditions, the depth of mining carried out and surface conditions. Aquifers, watercourses on the surface of the ground, provide the potential for direct increase, uncontrolled inflow or water intrusion into the excavation.

The climatic hazard is described by determining the so-called substitute climate temperature in the manner specified in the Polish Standard or by determining the microclimate parameters of the mine air. Using the climate proxy temperature, the working time of the mining personnel in the excavation is determined.

Here, the rocks or coal are rapidly moved/ejected by means of energy that have accumulated in the rock mass. Gas and rock blast hazards fall into three categories.

Fire hazards in underground mining are endogenous and exogenous. Endogenous hazards are caused by spontaneous heating of coal which is a natural phenomenon. Endogenous hazards are caused by human factors [14]. Table 1 presents the characteristics of the longwalls in terms of the analysis of mining and geological conditions in which the pressure monitoring system of the powered roof support units was used. Table 2 shows the list and data of the coal walls included in the pressure monitoring.

Table 1. Characteristics of natural hazards in longwalls subject to pressure monitoring.

longwall	Type of hazard						
	methane [category]	rock bursts [degree]	coal dust explosion	water [degree]	atmospheric	gas or rock outbursts	fire
393a	-	I	A	I-II	-	-	V
C-3	III	I	B	I	-	-	I, II, III
C-1	III	I	B	I	-	-	II, III
W-7	III	I	B	I	-	-	II
3b-S	IV	II	B	I	I	-	III
TB103	IV	I	B	I	-	-	II
7a	IV	-	B	I	-	-	II

The compilation of data makes it possible, in the first stage of selection of the support system, to analyse prevailing conditions in selected longwalls before introducing the pressure monitoring system. The data collected on mining and geological conditions are used to select the type and scope of roof support operation and pressure monitoring system that will be used in the longwall units.

Table 2. List of analysed pressure monitoring at individual longwalls and data of coal faces subject to pressure monitoring.

Mine	Longwall	Seam	Panel length [m]	Length [m]	Type of roof support	No. of units in the longwall [pcs.]
Piast	393a	209	440	165	Glinik-21/46-POz	109
Marcel	C-3	507	448	180	Hydromel 16/41 POz	118
					Glinik 14/34 POz	3
Marcel	C-1	505wg	1,056	119-128	Glinik 14/34	89
Marcel	W-7	505	839	180-190	Hydromel 16/41 POz	113
					Glinik 14/34 POz	26
Murcki-Staszic	3b-S	510	921	155	KW-17/43 POzW1/ZRP	
					KW-17/43 POzW1/BSN/ZRP	103
Sośnica	TB103	414/2	895	233	ZRP-15/35-POz	130
					ZRP-15/35-POz/BSN	8
					KW-18/34-POzW1/ZRP	17
Halemba	7a	402/K	730	230	KW-16/32-Poz/ZRP	130

3. Mode of operation of the pressure monitoring system for powered roof support units

The system works in a form of wireless communication among pressure sensors located in units of the powered roof supports. Interruptions of selected sensors do not affect the continuity of data transmission. The sensors are equipped with a replaceable energy source in the form of a battery, whose average operating time is about 2 years. The battery is approved for replacement in underground conditions with regard to hazards without the need to dismantle the sensor. The data acquired by the sensors are sent to a station located in the main workings. This station is equipped with a radio system, a cable interface for communication with a computer. The main station can be used for 3D visualisation of pressure distribution in powered roof support units.

They can be used to determine the working load bearing capacity of the support unit. They also include the pressure values in the legs of the powered roof support, on the supply or supply bus, as well as the supply voltage values of the sensors, so that links replacements can be planned in advance.

IOP Conf. Series: Materials Science and Engineering 1134 (2021) 012003 doi:10.1088/1757-899X/1134/1/012003

The system is equipped with a signalling system mounted on the canopy of the powered roof support unit. The signalling allows, through three colours (red, green, orange), to determine the pressure state during the working phase of the unit. This feature makes it easier for employees to work. In addition, it increases efficiency, improves safety for workers and powered roof support sections.

The underground station uses optical fibre or a teletechnical network to transmit data to the surface. There is a control station on the surface where it is possible, via the website, keep of archives, view in 3D and analyse the data sent by the sensors operating in the longwall workings. The web access facility has the additional feature of allowing general access to the data of several users and viewing the operation of powered roof support units in several longwall workings. The 3D visualisation and data analysis enables an earlier detection of casing unit failures and the efficiency of units and workers.

The wireless pressure monitoring system, whose sensors are built into hydraulic jacks and the floor support cylinder, makes it possible, based on the data collected, to visualise the work cycle of a powered roof support unit, and to analyse the occurrence of leaks in the hydraulic systems and power hydraulics. The data extracted from the monitoring system forms an algorithm to represent the pressure state for the mechanised shoring unit, as shown in Figure 1.

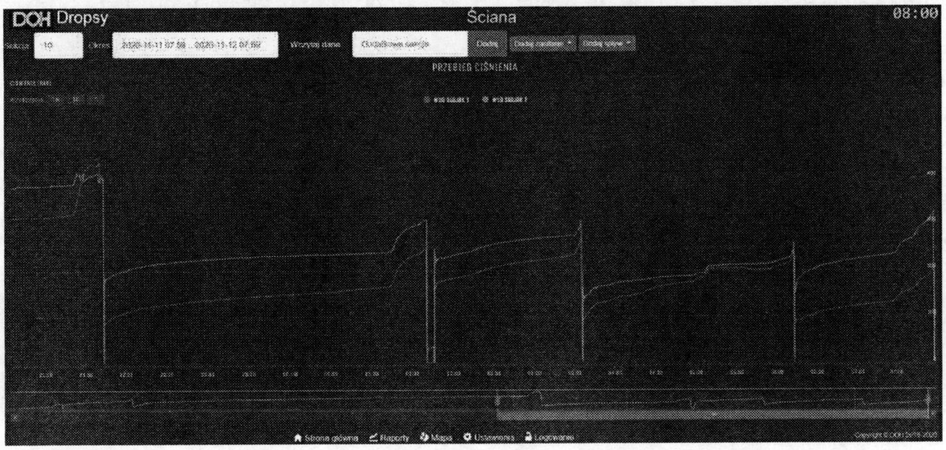

Figure 1. Example of a course of recorded pressures for a selected section of the powered roof support, where; the blue line represents the right leg, and the red line represents the left leg.

The devices included in wireless pressure monitoring are; a wireless pressure transmitter, whose task is to measure pressure every second when connected to the hydraulic system. Its task is to measure the pressure in the leg of the support unit. The sensors are also located in the supply and discharge bus, as standard at three spots in the longwall. In addition, sensors are placed in the floor base support actuator. The device, providing the connection between the system elements in the wireless network and the device communicating by wire, is an intrinsically safe radio converter. Its operation is characterised by the conversion of a wireless signal into a wired one. The current distribution of pressure in the longwall for the hydraulic legs, the supply and return lines and the floor base support actuator are displayed at the underground visualisation station. This is a computer in a flameproof case, which can also generate pressure distribution graphs for the entire longwall, as shown in Figure 2, and displays archives of the data.

IOP Conf. Series: Materials Science and Engineering 1134 (2021) 012003 doi:10.1088/1757-899X/1134/1/012003

Figure 2. Pressure distribution for powered roof support in a longwall, where: red represents insufficient spragging of the roof support, and green represents proper spragging of the roof support.

The transmission of the collected data between the excavation and the surface server is carried out using the optical fibre. The data, collected by the surface server, are made available to the dispatching station.

4. Surveys

Due to several activities, related to an improvement of occupational safety in hard coal mining, a preliminary survey was conducted among mine employees regarding the use of the pressure monitoring system for powered roof support units and an implementation of modern technologies in mining. The introduction of pressure monitoring systems in the mining sector received negative feedback from employees because of fear of the unknown and the cost of implementation. Given the vital importance of hard coal mining in the Polish economy and the need to improve efficiency, a preliminary survey of employees was carried out to address the technical barrier.

The survey has been designed for a group of 70 employees, working in the departments where investments are planned. This group consisted of physical workers and supervisors of an underground hard coal mining plant. The average age was 37 years and the average period of service was 13 years. Respondents were asked to answer the questions by marking an "X" in the appropriate box. The results are shown in Table 3 and in the graph (Figure 3).

Table 3. The survey designed to determine an impact of modern technology on the manner of performing tasks.

Item	Questions	Responses					
		yes	no	no opinion	yes	no	no opinion
1	Is the applied control system of the powered roof support unit easy to use?	62	3	5	89%	4%	7%
2	Is it reasonable to introduce a system for a visualisation of emergency states of operation of units (including assessment of supply pressure drop)?	68	1	1	97%	1%	1%
3	Should the control system have a warning signalling system during the switching off units (including a signalling system prohibiting entry under the unit being modified)?	58	7	5	83%	10%	7%

4	Is the message of the warning signal unambiguous/readable?	52	15	3	74%	21%	4%
5	Are you in favour of introducing new technologies for controlling powered roof support units?	67	2	1	96%	3%	1%
6	Have you received training on the correct use of a powered roof support unit equipped with the pressure monitoring system?	38	30	2	54%	43%	3%
7	Are you satisfied with the current control technology of the powered roof support unit?	55	15	0	79%	21%	0%
8	Is direct control of the support unit a good solution?	55	10	5	79%	14%	7%
9	Have you had the opportunity to work with an autonomous longwall unit?	0	70	0	0%	100 %	0%
10	Would an autonomous longwall complex improve safety during the ongoing work?	27	40	3	39%	57%	4%
11	Have you got any experience with electro-hydraulic control?	1	68	1	1%	97%	1%
12	Are you in favour of introducing electro-hydraulic control of powered roof supports?	5	52	13	7%	74%	19%
13	Are you in favour of introducing a single pressure monitoring system for powered roof support units across the entire company?	64	3	3	91%	4%	4%
14	Do you think that powered roof support is a complex machine?	3	67	0	4%	96%	0%
15	Are current IT systems made in an employee-friendly way?	62	3	5	89%	4%	7%
16	Are the currently available operational manuals for machinery and equipment in the mine written in a clear and comprehensive manner?	34	33	3	49%	47%	4%
17	Is it necessary to know the construction and principle of operation of powered roof support to operate it?	67	3	0	96%	4%	0%
18	Is training provided in the mine on the construction, operation and principle of operation of powered roof support units?	32	36	2	46%	51%	3%
19	If the answer to question 18 is YES, is the level of this training sufficient?	27	5	38	39%	7%	54%
20	If the answer to question 18 is NO, do you think such training should be provided?	68	0	2	97%	0%	3%
21	Does the pressure monitoring system improve occupational safety?	50	15	5	71%	21%	7%
22	Can you keep up with mastering the use of increasingly modern technical devices in everyday life?	57	13	0	81%	19%	0%

The preliminary analysis shows that the majority of employees support an implementation of modern technology and use of the pressure monitoring system for powered roof support units. The employees agree that innovative solutions improve safety and comfort at work. However, it is worth noting that as a result of reduced knowledge and a presentation of these solutions in an incomprehensive way, the employees feel anxious, mainly due to previous habits and traditions of their work.

IOP Conf. Series: Materials Science and Engineering 1134 (2021) 012003 doi:10.1088/1757-899X/1134/1/012003

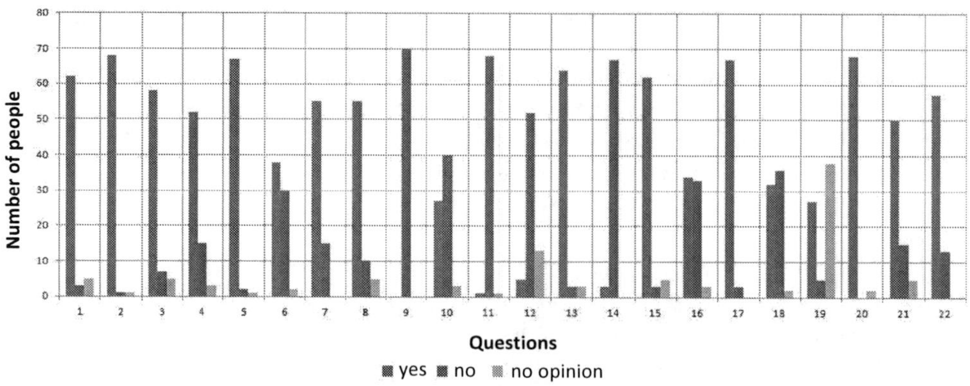

Figure 3. Responses of employees to survey questions.

The process of implementing modern solutions has increased dramatically, but there is still a long-term process before mining gets fully automated. If it does happen, it will take time for employees to change their attitudes, so more intensive trainings will enable to increase their awareness and reduce their fears. At present of technological development, it is essential to directionally improve the quality and frequency of courses and trainings provided. The development programmes for employees will enable them to expand their knowledge and experience in the safe use of modern solutions. Such activities are aimed at preparing employees and managers for market changes and a technical implementation of modern solutions following the ideas of Industry 4.0.

5. Conclusions

The Polish mining industry is characterised by the presence of many natural hazards. Therefore, it is important to evaluate occupational risks and identify preventive actions. The documents prepared before and during an exploitation of the deposit, should describe the procedure to be followed to ensure a safe operation of the mine, the occupational safety of employees, and a control of emergencies. Creating diagrams, technologies, a protection of buildings or environmental elements, makes it possible to visualise working conditions and a selection of machinery and equipment.

Each mining facility has individual mining and geological conditions. Therefore, with the worsening conditions in mining, it is inevitable to use the technologies proposed by the ideas of the fourth industrial revolution. One of the proposals of the fourth industrial revolution is flexible automation, which allows an introduction of programmable controllers and computers into industrial equipment. An example is the use of a monitoring system for powered roof support units in the mining industry. The monitoring system will make a significant contribution to improving workplace safety. Increased investments in machinery and equipment supplied with modern safety features and the quality of trainings, will enable employees to develop and feel comfortable at work. Aware and well-trained employees are the basis for limiting dangerous incidents and minimising their negative effects.

The presented scope of use of the pressure monitoring system of powered roof support units will provide an opportunity in the mining industry to:
– increase competitiveness in the mining market,
– increase safety,
– reduce the risk of losing sufficient unit support,
– prevent the occurrence of undesirable events,
– increase the efficiency of mining operations,
– increase productivity,
– store and analyse data,
– increase the effectiveness of the control and supervision of employees,
– detect and prevent failures earlier,
– indicate the differences in pressure values in a particular unit.

IOP Conf. Series: Materials Science and Engineering 1134 (2021) 012003 doi:10.1088/1757-899X/1134/1/012003

In addition, the paper presents a survey conducted among employees working in the mining industry. Its results provided an insight into the employees' approach to a realisation of their tasks using modern technologies. The data indicate that they have never had such opportunities before. They expressed a sincere desire to learn more about the machines and equipment that are being developed to improve their work comfort. Incomplete knowledge that they receive makes them anxious and afraid of the unknown. To change their attitude towards modern technologies, it is necessary to provide them with appropriate trainings aimed at presenting the scope of work and the techniques of using modern machinery and equipment in a simple and understandable way. It seems reasonable that the possibility of a dangerous incident, involving the human factor, can be significantly reduced through appropriate investments and use of modern technologies.

References

[1] Prusek S, Rajwa S, Wrana A and Krzemień A 2016 *Assessment of roof fall risk in longwall coal mines*, "International Journal of Mining, Reclamation and Environment", pp. 1-17

[2] Szurgacz D 2019 *Zmechanizowana obudowa ścianowa w zmiennych warunkach górniczo-geologicznych*, Oficyna Wyd. Politechniki Wrocławskiej

[3] Bazan Ł, Diederichs R, Gaja W, Lubryka J, Ptak K, Zych K 2015 *Doświadczenie eksploatacyjne z wdrożenia systemu monitoringu ciśnienia EH-Presss Cater w kopalni soli,* Napęd i Sterowanie 3, s. 124-129

[4] Oset K and Juzwa J 2015 *Bezpieczeństwo pracy urządzeń hydraulicznej obudowy zmechanizowanej i osób w rejonie ściany wydobywczej na przykładzie innowacyjnego systemu* X-MAN, Napęd i Sterowanie, 12, s. 110-114

[5] Jasiulek D, Bartoszek S and Lubryka J 2019 *Efektywność wykorzystania i bezpieczeństwo techniczne górniczej obudowy zmechanizowanej PRASS III,* Maszyny Górnicze 1, s. 73-79

[6] Michalak D, Jaszczyk Ł, Rozmus M, Wołczyk W and Lesisz R 2015 *Metody kształtowania bezpieczeństwa pracy z użyciem technologii informatycznych*, Maszyny Górnicze 3, s. 97-104

[7] Szurgacz D and Brodny J 2018 *Role and importance of pressure monitoring for elektro-hydraulic control system designed for powered roof supports,* "Conference proceedings, Science and technologies in geology, Exploration and Mining", 18, 1.3; pp. 793-800

[8] Gil J, Kołodziej M, Szurgacz D and Stoiński K 2018 *Introduction of standardization of powered roof supports to increase production efficiency of Polska Grupa Górnicza S.A.*, "Mining – Informatics, Automation and Electrical Engineering", 4(536), pp. 33-39

[9] Krauze K, Rączka W and Stopka G 2019 *Project and test results of new solution for powered roof support for low seams,* "Mining – Informatics, Automation and Electrical Engineering", 1(537), pp. 29-34

[10] Wyższy Urząd Górniczy 2019 *Ocena stanu bezpieczeństwa pracy, ratownictwa górniczego oraz bezpieczeństwa powszechnego w związku z działalnością górniczo-geologiczną w 2018 roku (porównanie od roku 2014),* s. 9-31

[11] Burtan Z, Stasica J and Rak Z 2017 *Wpływ katastrofogennych zagrożeń naturalnych na bezpieczeństwo pracy w górnictwie węgla kamiennego w latach 2000-2016,* Zeszyty Naukowe Instytutu Gospodarki Surowcami Mineralnymi i Energią Polskiej Akademii Nauk 101, s. 7-18

[12] Rozporządzenie Ministra Środowiska z dnia 29.01.2013 *w sprawie zagrożeń naturalnych w zakładach górniczych*, Dz. U. z 2015.poz.1702, z późn. zm.

[13] Sobik L, Brodny J, Buyalich G and Strelnikov P 2020 *Analysis of methane hazard in longwall working equipped with a powered longwall complex*, E3S Web Conf. Volume 174, (01011).

[14] Szurgacz D, Sobik S and Brodny J 2019 Integrated Method of Reducing the Threat of Endogenous Fires in Hard Coal Mines, E3S Web Conf. Volume 105, (01013)

IOP Conf. Series: Materials Science and Engineering 1134 (2021) 012004 doi:10.1088/1757-899X/1134/1/012004

Innovative Concept of Production Support System for the LW Bogdanka Mine

A Dyczko[1], M Malec[2,3]

[1] Mineral and Energy Economy Research Institute of the Polish Academy of Sciences, J. Wybickiego 7a, 31-261 Kraków, Poland
[2] KOMAG Institute of Mining Technology, Pszczyńska 37, 44-101 Gliwice, Poland

[3] E-mail: mmalec@komag.eu

Abstract. The research work results, presented in the article, describe a system concerning the management of a production line in the aspect of stabilization and improvement of the run-of-mine quality and maximization of economic effects. Basing on the geological deposit model, the authors concentrated on designing mining operations and on planning production parameters. Special attention was paid to a development of the mine workings design module, the mining operations scheduling module, the module for mixing and optimization from the quality point of view as well as the module for mining operations revenues and forecasts of costs. Forecasting the production quality, planning and integration of mining processes with coal preparation processes and sales are discussed in detail. The article is ended with some recommendations of general character, including quality management at the stage of mining and production planning, quality monitoring at the stage of coal mining, and its preparation. The system is oriented onto production quality forecasting and onto an integration of mining, preparation and sales planning.

1. Introduction

An automation of production processes, implemented in the LW Bogdanka, required studies and analyses of several interconnected components. A development of an innovative concept, enabling the management of the production line in the aspect of stabilization and improvement of the run-of-mine quality and maximization of economic effects was possible due to a realization of some research projects at the KGHM Polska Miedź S.A., JSW-Jastrzębska Coal Company S.A. and the TAURON Wydobycie S.A. The developed concept is based on the Integration Platform which is a crucial component of the presented solution. As a matter of fact it is a good and convenient tool for an integration of different pieces of information, enabling a standardized exchange of data among systems built in various technologies, using various communication protocols.

An innovative approach to the management of the production line consists in a division of a deposit as regards its quality, including mineralogical and petrographic features to establish quality parameters for a preparation of minerals. In the case of hard coal this division is based on physical properties (a division into coal grades) and also on the ash and sulphur content. Coal quality has a decisive impact on its further use. The division criteria must be clearly defined [1]. If a few different mineral grades occur in the deposit, then their distribution is presented on maps and cross-sections and their amount must be calculated separately. Qualitative and quantitative features of the deposit have an impact on its economic value. The following factors decide about the economic value of the deposit:

IOP Conf. Series: Materials Science and Engineering 1134 (2021) 012004 doi:10.1088/1757-899X/1134/1/012004

- a deposit attractiveness for management determined by its depth, structure, configuration and mineral properties,
- a degree of the deposit development for management,
- a suggested method of deposit management,
- a risk related to an assessment of technical and economic conditions of mining and preparation processes as well as sales of products obtained from the minerals' production line.

The aforementioned factors are interrelated, but it is important to remember that they are either related to the natural properties of the deposit or to the activities oriented onto its management. A design of the deposit management together with technical and economic analyses of the plant construction form the basis for a determination of commercial resources. These analyses clarify and justify the following management components:

- a choice of the deposit opening method and of its mining method,
- a determination of basic mining-and-geological parameters,
- a determination of mining and preparation costs,
- a determination of economic factors.

Figure 1 presents a classification of resources from the point of view of their economic importance.

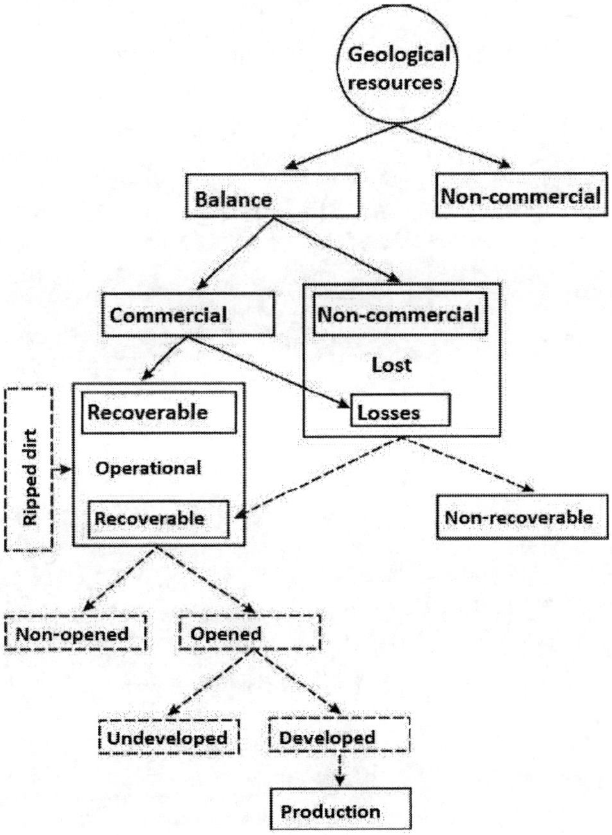

Figure 1. Classification of resources in terms of their economic importance [2].

The mining losses (quantitative losses) substantially reduce the deposit mining efficiency, resulting in increased expenditure on development work. The dilution (qualitative losses) causes transport and preparation costs increase. Big thickness changeability, forcing to rip the dirt to obtain the required cutting height and also a big number of dirt bands, result in increased amount of dilution which rises costs of preparation processes. Due to an appropriate selection of mining systems and machinery parameters, it is possible to reduce the amount of dirt and thus to improve economic effects. Analyzing a dilution of ore deposits some researchers [3]-[5] concentrated on the factors which cause this phenomenon. They distinguished two types of dilution such as:

- a mining dilution resulting directly from the carried out mining operations where the dirt, situated close to the mined mineral, crumbles into pieces,
- a structural dilution, related to the nature of the deposit itself, results from a presence of individual spoil formations within the deposit.

Noppe [6] expressed an opinion that in the case of longwall mining, when cutting is carried out at the full seam height, it is not possible to avoid dirt in the run-of-mine. The sources of dilution can be divided into three groups:

- a primary dilution originating from ripping the seam roof or floor,
- a secondary dilution resulting from ripping the floor uplift or from the roof fall,
- a third-rate dilution caused by cleaning powered roof support units or mixing the dirt from drifts with the coal run-of-mine from longwalls.

In the case of underground mines Saeedi at al. [7] ranges so-called off-seam dilution (OSD) from 15 to 30% in relation to mining-and-geological conditions as well as to technical parameters of the machinery used in a given mining system.
Sources of dilution in a longwall face are presented in Figure 2.

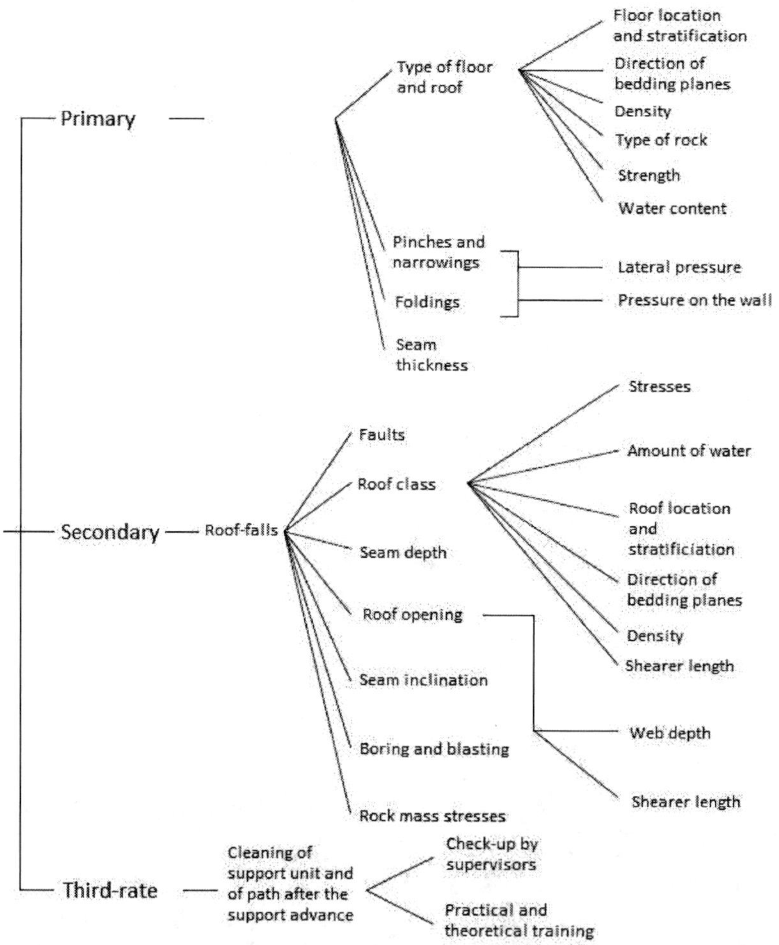

Figure 2. Sources of dilution in a longwall face [6].

2. Research objectives

The research methods and tools, applied for a development of an innovative system for the LW Bogdanka, were used for confirming the fact that it is possible to achieve positive economic results of rationalizing the coal winning processes consisting in extracting less dirt, where it is geologically and technically justified. Technical and economic analysis should include basic elements which are decisive in the management of the coal production line, in particular:

- a selection of a deposit opening method and of a deposit exploitation system [8, 9],
- a determination of basic mining-and-geological parameters [10, 11],
- a determination of costs of exploitation and beneficiation processes [12, 13],
- a determination of economic indicators.

An implementation of a correct mining system and of correctly selected machinery reduces losses and impoverishment significantly. The research methods, techniques and tools presented in the article include:

- an expert assessment of Polish and foreign literature [14, 15, 16],
- mathematical modelling,
- probabilistic modelling,
- statistical analysis,
- assessment of economic efficiency [17].

3. Concept of production support system

The innovative concept of production support system is based on a holistic and integrated approach to the movement of machinery and people, flow of production, transport and service tasks in terms of production processes planning, organizing, controlling and inspecting. Basic tasks of the system, to be used in an integrated information environment of underground mines, comprise:

- designing and scheduling of mining projects based on a numerical deposit model together with an application of parametric methods for designing mine workings, transport systems and the infrastructure, forecasting production rates and costs as well as with an optimization of economic parameters,
- planning and organizing the production line, service, transport including employees, equipment, machines and materials,
- coordination and control of production, transport and service tasks as well as a circulation of documents,
- management of full life cycles of production projects and equipment,
- control of the production quality,
- integration, gathering, processing and reporting the data registered in the economic IT environment and technical systems,
- optimization of raw materials storage, transport and preparation based on algorithms and models of advanced process control (APC).

The concept of the production support system is oriented onto improving the main processes, identified within the numerical deposit model (NDM), such as:

- geological work designing and performance,
- interpretation and documentation of geological structure of the deposit together with its register of resources,
- support of mining and production planning,
- support of coal mining processes.

An implementation of a specialized tool (CAD) to create a numerical deposit model is the recommended solution. Linking of the Geological Database (GDB) and the Geological Deposit Model (GDM) with the Central Data Warehouse enables a bidirectional exchange of information with the

IOP Conf. Series: Materials Science and Engineering 1134 (2021) 012004 doi:10.1088/1757-899X/1134/1/012004

Map Data (MAP) area, making the deposit model available to the Mining Operations Scheduling (MOS) area. Figure 3 shows a schematic diagram of the system structure.

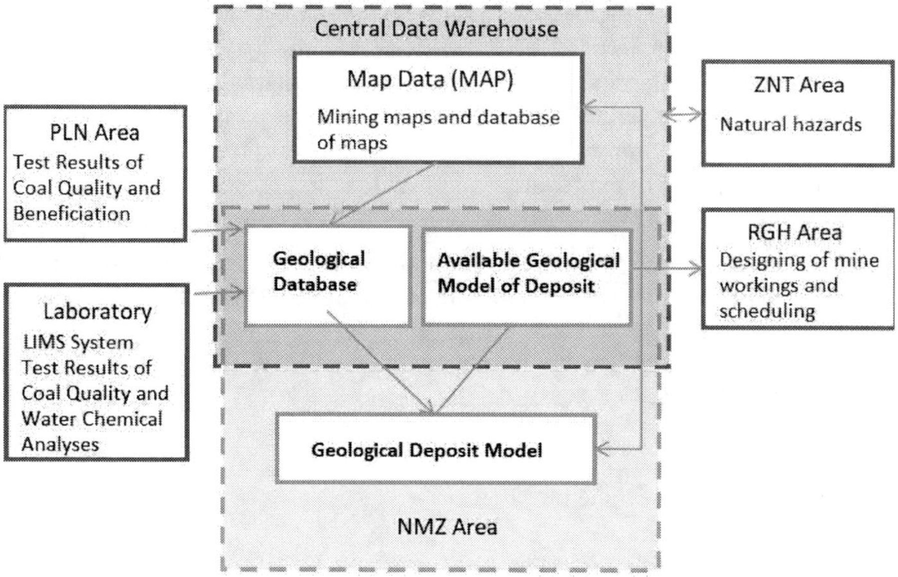

Figure 3. Diagram of the target solution in the numerical deposit model (NDM) area and its links with other areas of the production support system [18].

The Geological Database (GDB) is designed as an element of the Central Data Warehouse. Full geological information includes:

- lithographic profiles of surface boreholes, underground boreholes and mine workings,
- results of coal quality sampling (qualitative, chemical, physical and physiochemical data as well as data related to an analysis of coal washability),
- hydrogeological information containing measurements of the water table height, of the water inflow, of the water chemism and of underground water radioactivity,
- results of geomechanical tests in penetrometric boreholes.

The Geological Database (GDB) enables to record and service geological samples together with bidirectional integration with the Map Data (MAP) area.

4. Geological deposit model

Two methods were used for geological modelling: stratigraphic modelling and block modelling. Various sets of rules were used for different seams, enabling to develop separate models in various locations and in various ranges of the deposit stratigraphic profile. The outcome consisted of a stratigraphic table and a set of surfaces (grid), representing all the strata and stratigraphic surfaces selected for modelling. The data on seam qualitative parameters and dirt bands were processed in a similar way. As far as a development of grid models was concerned, several interpolator methods such as: finite element, inverse distance, kriging, nearest neighbour and triangulation methods were available.

The following data were processed:

- lithostratigraphic sequences,
- fault strike lines,
- profiles documenting exposures of lithological and stratigraphic strata,
- point findings of individual stratigraphic surfaces,
- graphical elements, imposed by the user, forcing certain geometrical parameters of the model.

Due to the tectonic structure complexity, modelling of faults is an important element of the deposit

model functionality. Faults are created in the system basing on point findings. The coal seams splitting is also modelled. The modelling of qualitative parameters (calorific value, sulphur content, ash content, moisture content, density etc.) is one of basic requirements. The developed qualitative models are used to calculate resources as well as for the quality planning processes.
Hydrogeological data are used in the deposit model to create thematic point maps. The data are used to develop models of the water table position in the selected water-bearing levels and to generate 3D visualization of water reservoirs.

Calculations of the deposit resources are carried out within the management system. Resource reports are produced on the current deposit model. The geological data and industrial elements of the model are visualized. The solution, presented in the concept, utilizes the Central Data Warehouse to report and make the required geological contents available. Visualizations are saved to .dwg files and they are also printed and exported to other data exchange formats e.g. pdf.

5. Mining operations designing and production planning (MOS)

The MOS system supports the process of mining operations designing and scheduling as well as the process of production scheduling with a possibility of quick updating, carrying out simulations and integrating a selected area with other components of the Production Support System. These processes are based on the geological deposit model, described in Item 4 of this article. The MOS area is a crucial activity in terms of short- and long- term planning of operational activity, planning of costs, investments, repairs etc.

The target process of production preparation and planning starts from the market analysis in terms of the possibility of selling individual products group. Figure 4 presents the diagram of the MOS system structure, containing four models:

- mine workings designing module,
- mining operations scheduling module,
- quality optimization module,
- module of revenues and costs forecasts of mining operations.

The process of mining operations designing, production planning and mining operations scheduling provides feedback to other management areas enabling to obtain final products of specified quality parameters. Mine workings designing module in the MOS enables a better understanding of the strata deposition and the course of planned workings (type, function, cross-section). It contains complete information on the deposit. It is integrated with the numerical deposit model. Mining operations scheduling module in the MOS plays an important role, as it enables to generate and analyze several variants aimed at finding the optimal solution in terms of quality and quantity of the products. In this case the speed of responding to changes in the deposit geological exploration is significant because it requires an update of designs and schedules. This module, combined with the preparation and sales support system, generates the data indispensable for forecasting revenues and costs of mining operations. The module for mixing and optimization, in terms of quality, is responsible for defining qualitative objectives in conjunction with quantitative objectives. This module provides a tool for scheduling mining operations. The MOS also includes the module for mining operations revenues and costs forecasting, which is closely related to the mining operations scheduling module. It provides the information on unit costs of individual mining operations and on the liabilities resulting from mining operations. Such an approach enables to obtain a complete and reliable assessment of economic-and-financial efficiency of a mining operation under analysis.

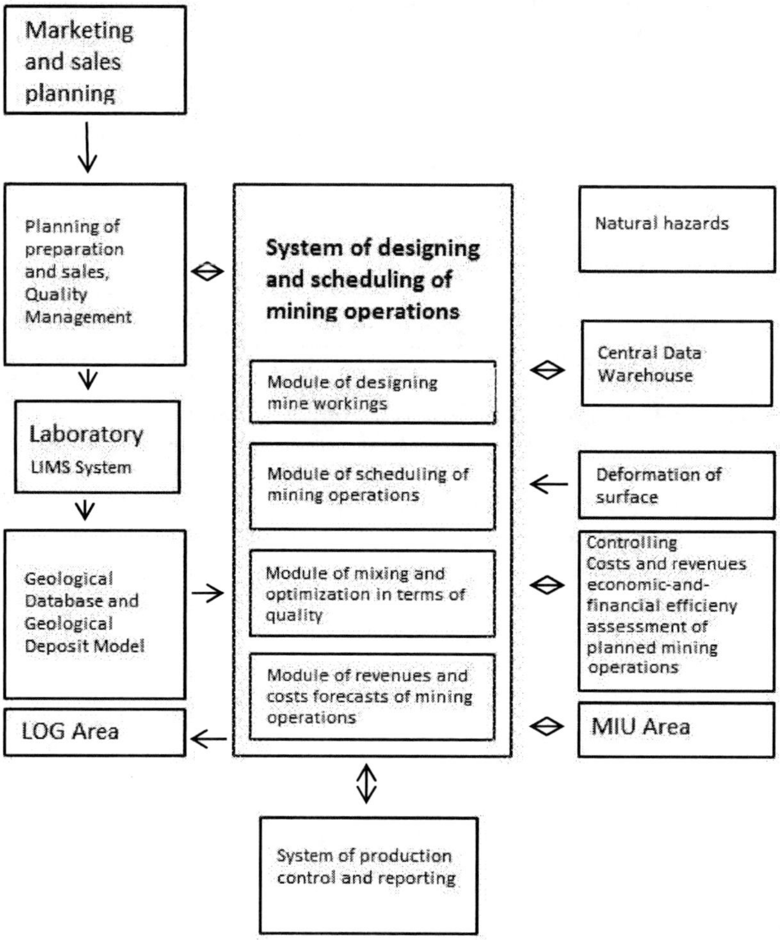

Figure 4. Diagram of production designing and scheduling system in terms of quality and links with other areas [18].

6. An integration of planning processes on the mining side with processes of coal preparation and sales

The concept of the process of a mining plant production planning and controlling, presented in the article, responds to the market demand and intended sales in individual product groups. Such an approach causes an urgent need of controlling the run-of-mine quality.

In accordance with the presented concept of the production line management the broadly understood quality area is linked to all the operational areas of a mine (Figure 5).

IOP Conf. Series: Materials Science and Engineering 1134 (2021) 012004 doi:10.1088/1757-899X/1134/1/012004

Figure 5. Links of quality planning are with other areas of mine operations [18].

The integration among individual elements of the production system is extremely important because of the system functionalities and the measuring infrastructure. The production reporting area is supplied with information on the quality of semi-products and products. The quality management, at the stage of mining and production planning, is closely related to the solution developed for mining operations scheduling. The preparation and sales planning is combined with the quality control system. In such an approach the forecast of raw coal quality is an integral part of the mining operations schedule. Developed variants of mining plans may be optimized in view of satisfying different objectives: qualitative, quantitative, cost etc. Presenting the management of the run-of-mine production line, it is worth highlighting quality monitoring at the stage of coal mining and its preparation. Some corrective and preventive measures should be implemented in the case of deviations from the planned values. The system enables continuous monitoring of coal quality parameters. Results of all the measurements are transferred to the SCADA system and then to the acquisition platform of the data repository. A quality controller function relies on analyzers installed on belt conveyors. Due to such a function it is possible to control a realization of the quality plan, being an inseparable part of the production schedules. The recommended solution results in reducing the frequency of manual sampling in favour of using a continuous quality measurement system and making the results available in the production reporting system.

7. System for production quality forecasting enabling integration of mining, preparation and sales planning

It is an automatic system enabling to gather data from sensors installed on belt conveyors in various places of the run-of-mine haulage from the longwalls and from the preparation plant. Figure 6 presents the components of this system and their links with other production areas under analysis.

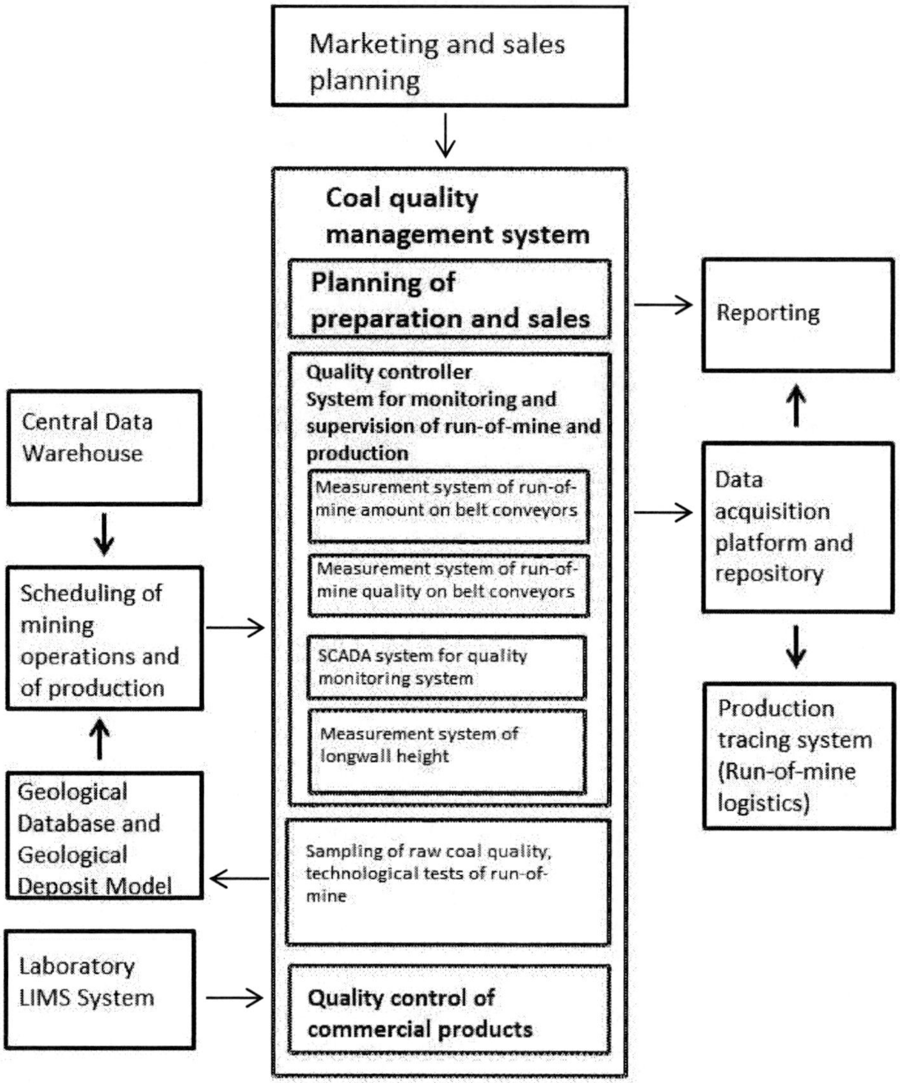

Figure 6. Diagram of coal quality management and its links with other production areas [18].

The quality management at the stage of preparation and sales planning specifies the required quantities and qualities of raw coal to meet the production assumptions related to commercial products. These requirements are transferred to the mining operations designing and scheduling system. The SCADA system collects the information from individual quality parameters monitoring devices and visualizes it in real time. The last stage of the quality control process consists in taking qualitative samples of saleable coal. Quality control tasks comprise monitoring of the production quality and quantity to ensure the quality level consistent with the commercial offer. The implementation of the coal production line management system may require an expansion of the existing IT network at the LW Bogdanka Mine, including the SCADA system as well as a purchase or a modernization of workstations.

8. Summary and conclusions

At present the coal mining sector experiences serious problems due to rather low prices and due to the European Union policy oriented onto a decarbonization, so it is important to maintain high flexibility of each mine in the context of products range and production costs. A pragmatic approach requires an implementation of adaptive planning mechanisms and techniques aiming at optimizing the operational activities. A capital intensive nature of mining processes, combined with long planning horizons and a big inertia of the production system, make a quick reaction to changing coal prices very difficult, because each scenario must be planned in advance. In many cases it is more a kind of a forecast than an exactly scheduled plan. It also concerns the methods of operational activity improvement, which is a long-lasting process. In this context a continuous optimization of long- and short-term plans becomes a must [19].

In general, the production support system, presented in the article, like any IT solution, is a tool enabling to facilitate and improve the management of the coal production line suggested for the LW Bogdanka. An implementation of a full PDCA (Plan-Do-Check-Act) cycle, based on continuous gathering of data, provides measurable effects. The concept of the IT system assumes an automatic calculation of the quantity and quality of the run-of-mine as well as of the dirt during selected time intervals. After a completion of simulation processes an automatic forecast is generated. Basing on the example of the LW Bogdanka, it was possible to determine an impact of qualitative losses on the efficiency of the coal mining operations. Economic effects undoubtedly depend on qualitative and quantitative criteria. Positive economic results can be reached at a relatively small cost, if the coal production support system is implemented. The Integration Platform is a central element of the system, being a reliable tool of information integration, enabling a standardized exchange of data among systems built in various technologies or using various communication protocols. The Integration Platform, suggested for the LW Bogdanka Mine, uses the Service Oriented Architecture (SOA) technology. The Technical Data Warehouse is a component of the Integration Platform which gathers the data from industrial automation systems. The data transfer from the control systems to the Technical Data Warehouse, by means of the Open Platform Communications Unified Architecture (OPC UA) standard, is a key component of the system. However, the most important component of the solution is the Support System for Mine Operations Maintenance. An effective analysis of information, gathered in the system, can substantially improve the quality of management decisions. It is recommended to build the Centre for Advanced Data Analysis (CADA). The mechanization and automation degree of basic processes, carried out in the mine, forces a necessity of combining production management and reporting areas with the area of machinery and equipment maintenance. This operation is fundamental from the point of view of the Total Productive Maintenance (TPM). An implementation of mechanisms, making the data available from the SCADA environment to the Manufacturing Execution System (MES), Computerized Maintenance Management Systems (CMMS) etc. are crucial factors that condition the entire solution effectiveness. The concept of building a decision support system in the field of production activity, using IT solutions and production monitoring, is one of the LW Bogdanka strategic objectives oriented onto building the Intelligent Solutions Mine.

References

[1] Nieć M, Górecki J, Sermet E, Szczepiński J, Ślusarczyk G, & Borowicz A 2015 Methodology of exploration and documentation of deposits and geological survey in mines - XVI Seminary in relation to III Polish Mining Congress. Górnictwo Odkrywkowe, 56 (in Polish)
[2] Nieć M 1982 Mining Plant Geology, Geological Publishers. Warsaw (in Polish)
[3] Ingler D 1984 Mining Methods, Rock Dilution and Ore Losses
[4] Knissel W, Jurgen H and Fahlbusch M 1995 Significance of Selective Mining Exploitation for Economical and Environmentally Beneficial Underground Ore Mining. International Mineral Resources Engineering, Imperial College Press 5, 2 pp. 165-174
[5] Wright E A 1983 Dilution and mining recovery-a review of the fundamentals
[6] Noppe M 2003 The measurement and control of dilution in an underground coal operation. In Proceedings of the 5th mining geology conference. Bendigo, pp. 189-195
[7] Saeedi G, Rezai B, Shareiar K & Oraee K 2008. Quantifying level of out-of-seam dilution for longwall mining method and its impact on yield of coal washing plant in Tabas coal mine. In Proceedings of the international seminar on mineral processing technology, Trivandrum, India, pp. 370-73

[8] Dyczko A 1998 Current economic situation and prospects for development of KGHM Polish Copper. AGH Quarterly: Mining", 22nd year, Book 2 (in Polish)
[9] Dyczko A 2002 Technical and economical aspect of copper ore dilution in the case of the mines operated by KGHM Polska Miedź SA. Gospodarka Surowcami Mineralnymi. Volume 18. Special Edition. Cracow (in Polish)
[10] Dyczko A 2004 Some Aspects of the Run-of-Mine Impoverishment in Copper Ore Mines. Conference Proceedings of the Management of Minerals. Publishers: IGSMiE PAN, Cracow (in Polish)
[11] Dziura T & Zapotocki A 1976 Tests of Mining Use of Copper Ore Deposit in the LGOM Mines, Zeszyt Naukowy Politechniki Śląskiej, No. 70, Gliwice (in Polish)
[12] Krasiczyński K 1974 Method for Calculating Ore and Metal Losses in Exploitational Processes. Cuprum No. 2. Wrocław (in Polish)
[13] Wanielista K 1979 Economic Assessment of Exploitational Losses and of Ore Impoverishment. Rudy I Metale Nieżelazne No. 9 (in Polish)
[14] Hurysz J & Sikora W 1963 Impact of Technical Parameters of Roof Caving Longwall on Production Concentration. Przegląd Górniczy No. 9. Katowice (in Polish)
[15] Rabsztyn J & Kozdrój M 1967 Mechanization of Coal Cutting and Loading. Śląsk Publishers. Katowice (in Polish)
[16] Walisko J 1962 Advantageous Length of Mechanized Longwalls. Przegląd Górniczy No. 12 (in Polish)
[17] Sitko W 1976 Economic of Mining. Publications of the Silesian Technical University No. 596. Gliwice (in Polish)
[18] Dyczko A 2018 Methodology for the assessment of the impact of out of seam dilution on the efficiency of hard coal production on the example of LW Bogdanka S.A. Doctoral thesis under scientific supervision of Prof. Roman Magda. AGH University of Science and Technology. Cracow (in Polish)
[19] Lisowski A 2013 Hard Coal Mining in Poland: a Critical Assessment of Conducting the Market Transformation and a Discussion of Problems Still to be Solved 2006-2013. Publishers: Central Mining Institute, Katowice (in Polish)

Tests of electric properties of supply-control system of mining floor-loader

A Niedworok[1,2], K Lesiak[1], Ł Orzech[1], M Talarek[1]

[1] KOMAG Institute of Mining Technology, Gliwice, Poland

[2] E-mail: aniedworok@komag.eu

Abstract. Tests of electric properties of the supply-control system designed for a floor-loader operated in underground mine workings, are presented in the article. The tests included functionality of the machine power electronics systems, consisting of the inverter controlling the induction motor and the charger of lithium-ion battery. The tests were concentrated on a verification of functional correctness of the system managing an operation of the battery (BMS) and its collaboration with the superordinated floor-loader control system. The obtained test results enabled to verify the accepted technical and functional assumptions of the supply-control system as well as to confirm a possibility of its safe operation in underground conditions (including the zones where an explosion hazard occurs). The supply-control system of the floor-loader is developed within the framework of the HYDKOM 75 project number POIR.04.01.02-00-0102/16, co-financed by the European Regional Development Fund.

1. Introduction

A floor-loader is a basic auxiliary machine, in most cases mounted on a caterpillar chassis, designed for a removal of waste rock and coal from roof and side walls (left after firing) [1, 2, 3]. An additional function of this type machines includes operations connected with pulling up of components being the mine infrastructure (e.g. pipes or rails).

In Figure 1 the floor-loader of BH3000 type produced by HYDROTECH J.S.C., is presented. At present the machines of this type are equipped with a wire electro-hydraulic drive. A concept of wire supply has a few disadvantages, among them the superordinated ones include: a limited mobility and a feeder cable susceptibility to mechanical damages. An attempt of eliminating the above-mentioned disadvantages consists in equipping the floor-loader with an autonomous source of supply, e.g. in a form of highly efficient battery. Such activities were undertaken within the framework of the project entitled "Innovative mobile machine with a universal electric drive system increasing the technical safety level" HYDKOM 75, realized by the consortium consisting of the KOMAG Institute of Mining Technology and the industrial partner HYDROTECH J.S.C. within the framework of the Operational Programme Intelligent Development 2014 - 2020. The HYDKOM 75 project assumes designing, manufacture and testing of the battery-operated floor-loader to be used in underground mine workings, where potential hazards of methane or coal dust explosions occur [4]. The test of electric properties of the supply-control system of the designed floor-loader in the aspect of functional check-ups of power electronics systems and of the battery management system (BMS) are discussed in the article [5].

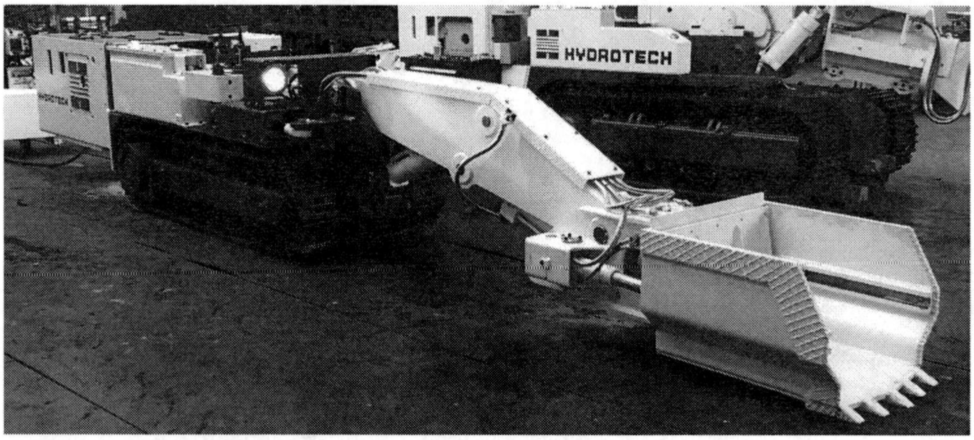

Figure 1. BH-3000 floor-loader made by HYDROTECH J.S.C.

2. Object and scope of tests

The object of tests includes a prototype supply-control system of the floor-loader, which can be divided into four blocks:

- a battery together with the management system (BMS),
- an inverter controlling an induction motor,
- a DC charger,
- a system of control and protections.

The system was equipped with 3-phase induction motor of 55 kW power to meet the testing needs. Electrical parameters of the used motor complied with the parameters of the motor which will be finally installed in the floor-loader under designing. The motor was supplied from the lithium battery assembly via the power electronics inverter. The batteries were charged by means of a DC charger directly supplied from the 500 V AC grid. Similarly as in the case of the motor, the battery pack had the same parameters as the pack to be used in the final version of the floor-loader. The IGBT transistors of power electronics equipment (i.e. the inverter and the charger) were cooled with water.

The main objective of the conducted tests was a check-up of the described supply-control system from the verification point of view of the accepted design assumptions, functional properties and possibilities of operating in the zones, where explosion hazards occur. In relation to that the following scope of tests was accepted and it included as follows:

- a determination of the correctness of cells' charging process,
- a determination of the proper realization of a driving operation (a process of discharging cells),
- a determination of the correctness of the cell battery management system,
- a determination of surface maximal temperatures,
- a determination of limit load for which a motor startup is correct.

The system under testing was installed on the motor test bench at KOMAG Institute of Mining Technology. The motor test stand of the power 223 kW enabled a free conditions' shaping of the induction motor supply-control system. In the system of the drive transmission a torque transducer, enabling a measurement of the actual torque and of the actual rotational speed was installed. The parameters of the electric system were measured by means of power analyzer. In Figure 2 a block diagram of the supply-control system of the floor-loader together with the measurement apparatus is presented.

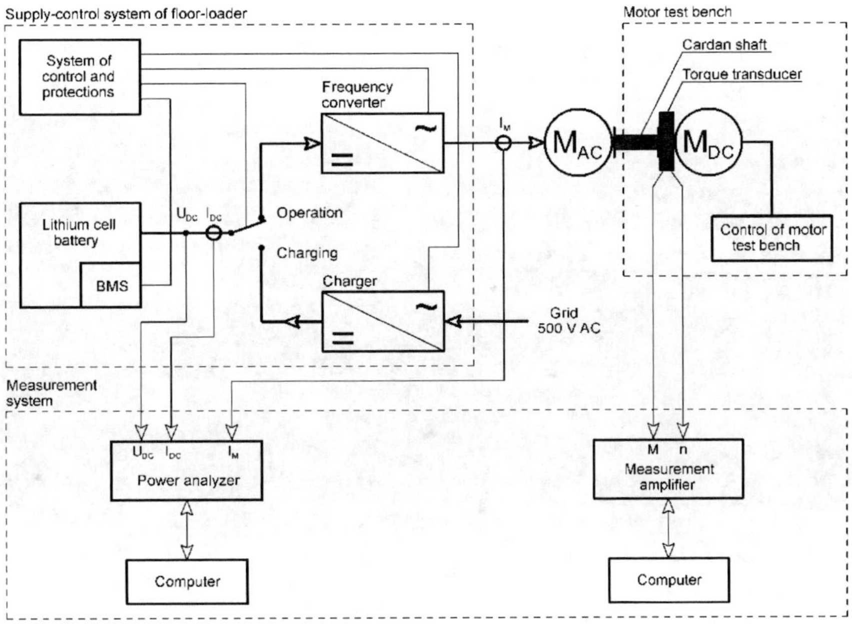

Figure 2. Block diagram of supply-control system of floor-loader together with measurement instrumentation.

Such a measurement system enabled taking measurements and recording of the following parameters:

- voltage in the battery DC circuit V_{DC},
- current in the battery DC circuit I_{DC},
- power in the battery DC circuit P_{DC},
- phase current on the output of the inverter (motor phase current) I_M,
- torque of the induction motor M,
- rotational speed of the induction motor n.

Simultaneously in the selected points of the system temperature measurements (among others: of the cassettes with battery and copper plates as well as of inverter and charger housings) were taken. In Figures 3 and 4 views of the supply-control system on the test stand are shown.

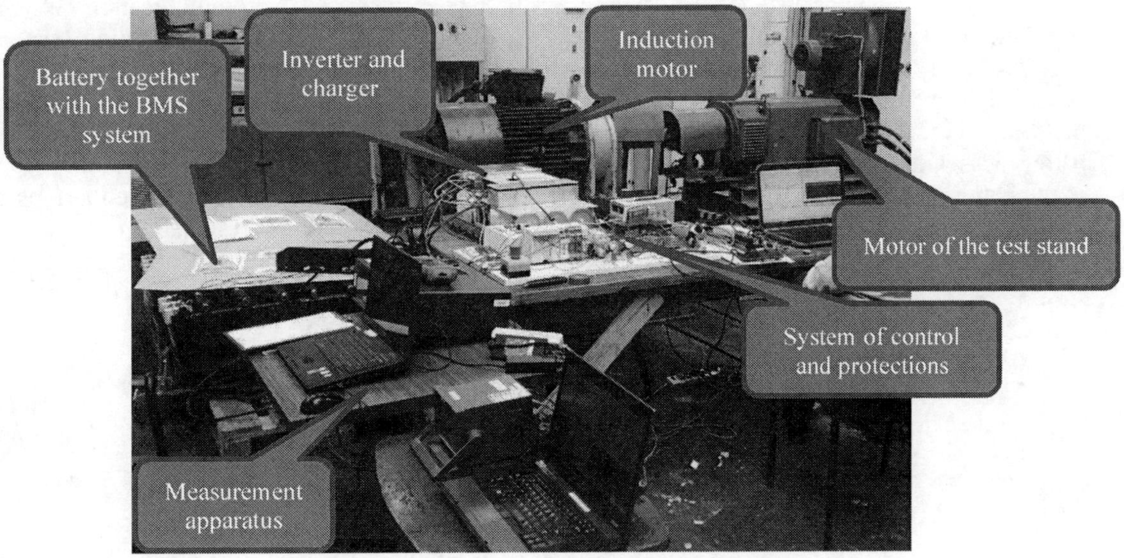

Figure 3. Supply-control system of the floor-loader on the test stand.

Figure 4. Inverter (left) and DC charger (right) installed on the water-cooled radiator.

3. Test course and test results

The objective of the test programme required to perform battery charging cycles and drive cycles (discharging the battery) of the supply-control system of the mining floor-loader. In the same time correctness of the cells' charging process, drive operation, operation of the BMS system were verified and the temperatures of selected system surfaces were measured.

3.1. Battery charging

The process of battery charging was realized by means of a charger supplied from a public distribution network 500 V AC. The charging current was 30 A. The charging process was controlled by the master control system, communicating with the loader by means of the CAN bus. A check-up of the charge level was realized by the management system (BMS) installed in the battery. Each charging cycle was started for the battery - discharged to the minimal level foreseen by the producer. Basing on the conducted tests minimal and maximal charging time which was about 2 to 3 hours [6] was determined. During the test some instantaneous charging breaks lasting about 4 seconds were observed. Their occurrence was caused by the breaks in the CAN communication between the superordinated control system and charger. The maximal battery voltage, after charging completion, was 766 V. The charging process ended automatically when the voltage on the battery single cell achieved the maximal value of 3.65 V. During the charging process maximal temperatures of the copper plate and of the charger housing were 35.2 °C and 37.4 °C respectively, however the temperature of the battery was 41.6 °C [6]. In Figure 5 exemplary time graphs of V_{DC} voltage, I_{DC} current and P_{DC} power in the battery circuit, recorded during the charging process, are shown. In Figure 6 the thermogram of the charger, made during charging by means of a thermovision camera, is presented.

IOP Conf. Series: Materials Science and Engineering 1134 (2021) 012005 doi:10.1088/1757-899X/1134/1/012005

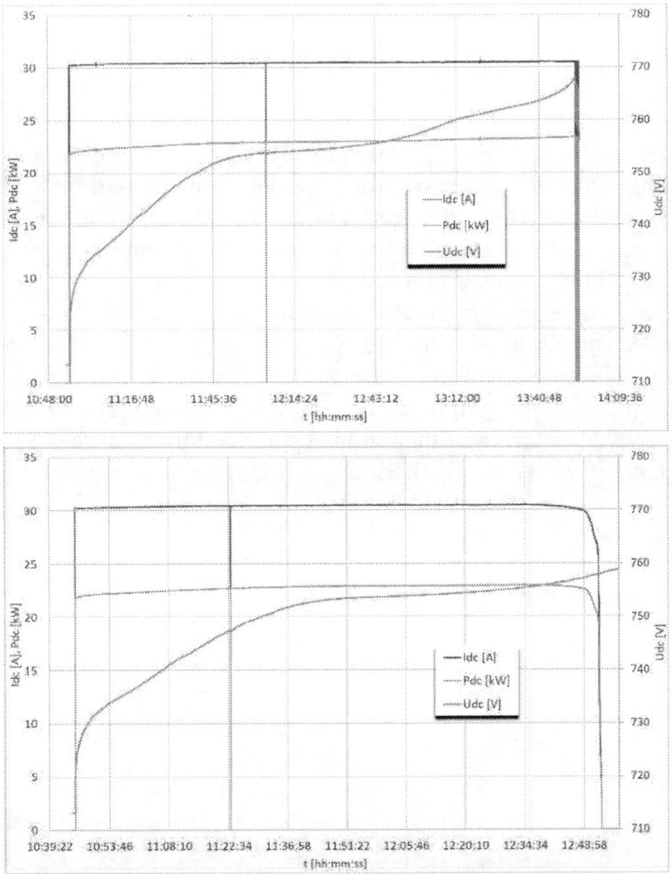

Figure 5. Exemplary time graphs of electrical parameters during the battery charging process.

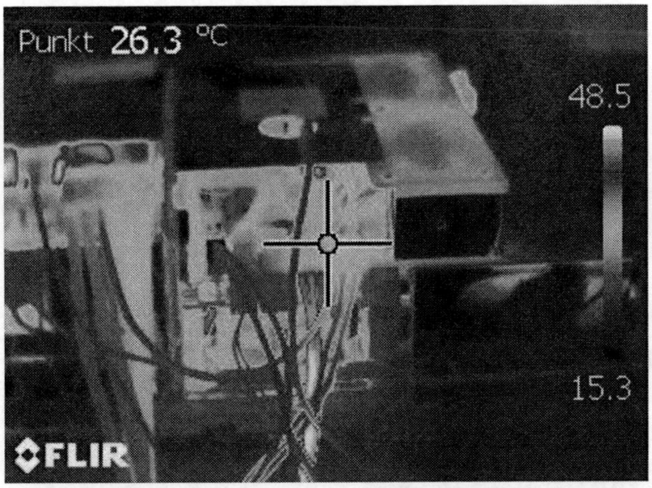

Figure 6. Thermogram of the charger made during the charging process (current of 30 A).

IOP Conf. Series: Materials Science and Engineering 1134 (2021) 012005 doi:10.1088/1757-899X/1134/1/012005

3.2. Drive operation (battery discharging)

The drive operation was realized by the frequency converter supplied from the battery of cells. The frequency converter supplied a three-phase induction motor of 55 kW power. The motor test bench was used for controlling the motor load to force constant, rated phase current of the frequency converter in the intensity of ~75 A. During the tests a constant torque and a constant rotational speed resulting from the applied frequency converter operational frequency (50 Hz) were maintained. Each of the test started after earlier charging the battery to the maximal level foreseen by the producer of cells. Basing on the conducted tests minimal and maximal operational time, which varied form about 57 minutes to about 1 hour and 6 minutes [1], was determined. The drive was switched off automatically when the voltage on the battery individual cell reached the minimal value of 2.65 V. During the driving mode operation maximal temperatures of the copper plate and of the frequency converter housing reached 27.5 °C and 31.1 °C respectively. However, the temperature of the cell battery was 48.0 °C [6]. In Figure 7 exemplary time graphs of the V_{DC} voltage, I_{DC} current and P_{DC} power in the battery circuit and the I_M phase current of the frequency converter are shown. In Figure 8 the frequency converter thermogram, made with use of a thermovision camera, is presented.

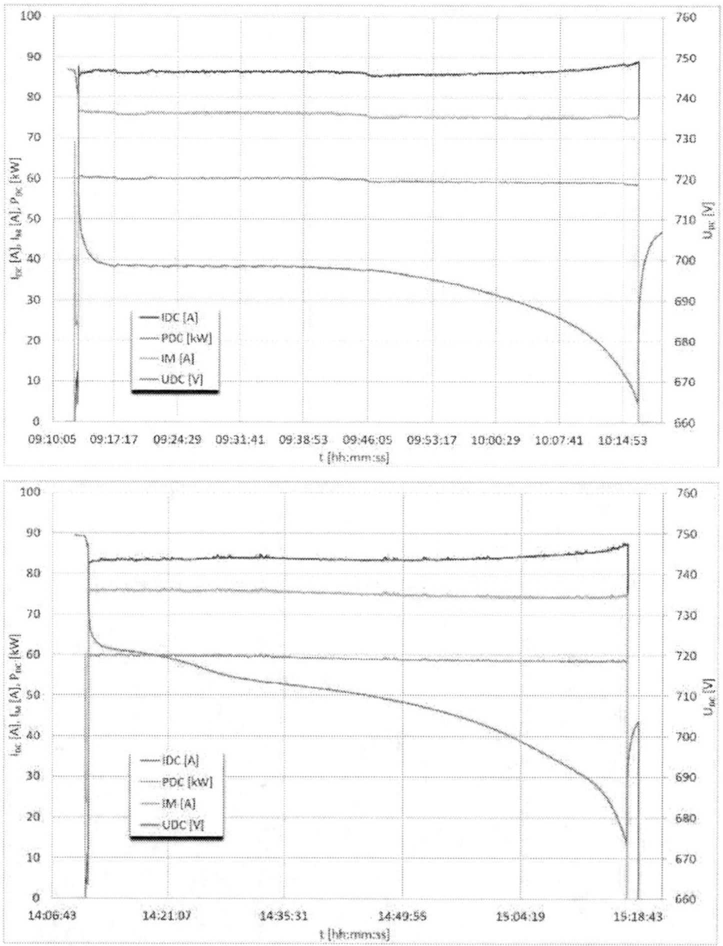

Figure 7. Exemplary time graphs of electrical parameters during the drive operation.

IOP Conf. Series: Materials Science and Engineering 1134 (2021) 012005 doi:10.1088/1757-899X/1134/1/012005

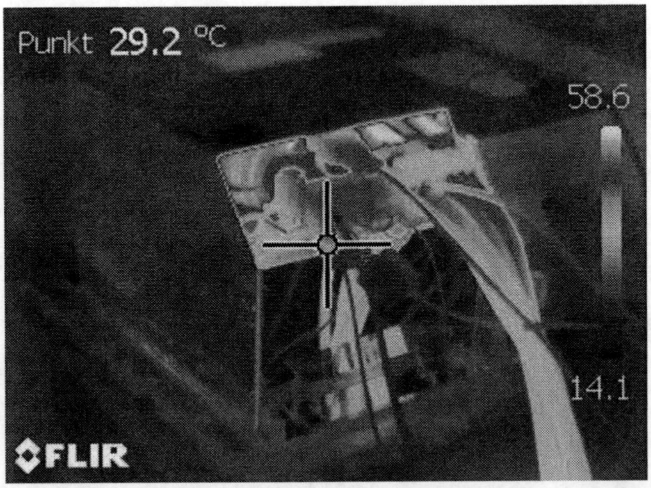

Figure 8. Thermogram of the frequency converter made during the drive operation (phase current ~75 A).

The tests were repeated for a forced half of the frequency converter rated phase current, i.e. 37.5 A. In this case the drive operation time was 2 hours and 39 minutes. For a half of the rated phase current lower values of maximal temperatures of the copper plate and the frequency converter housing, which were 22.7 °C and 27.3 °C respectively, were noticed. However, the temperature of the battery was 40.3 °C [6]. In Figure 9, time graphs of electrical parameters, recorded during the tests, are presented.

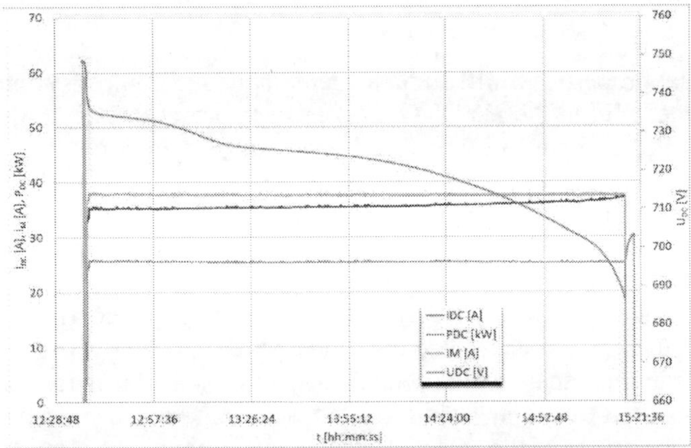

Figure 9. Exemplary time graphs of electrical parameters during the drive operation (phase current 37.5 A).

3.3. Motor startup tests

The tests aimed at a determination of the motor load value of the floor-loader supply-control system, for which a correct startup to the rated frequency (50 Hz) was possible. The motor load was established on the percentage value of the inverter rated current value (I_M), controlled by the motor test bench during its operation at the rated frequency. The tests were conducted for seven current values, i.e. 50%I_M, 60%I_M, 70%I_M, 80%I_M, 90%I_M, 95%I_M and 100%I_M. The load value, determined earlier by means of the motor test bench, was maintained during the drive startup. A condition of the correct startup consisted in starting the drive in such a way so that to "rotate" the motor to the rated speed. The tests were conducted on the charged battery. Basing on the obtained results, the load limit value, at which a correct motor startup was possible, reached 95%I_M. Above this value so called stall of the motor occurred, resulting in failure conditions of the inverter. In Figure 10 exemplary time graphs of mechanical and electrical parameters during the startup in the case of 95% value of the rated current, are presented.

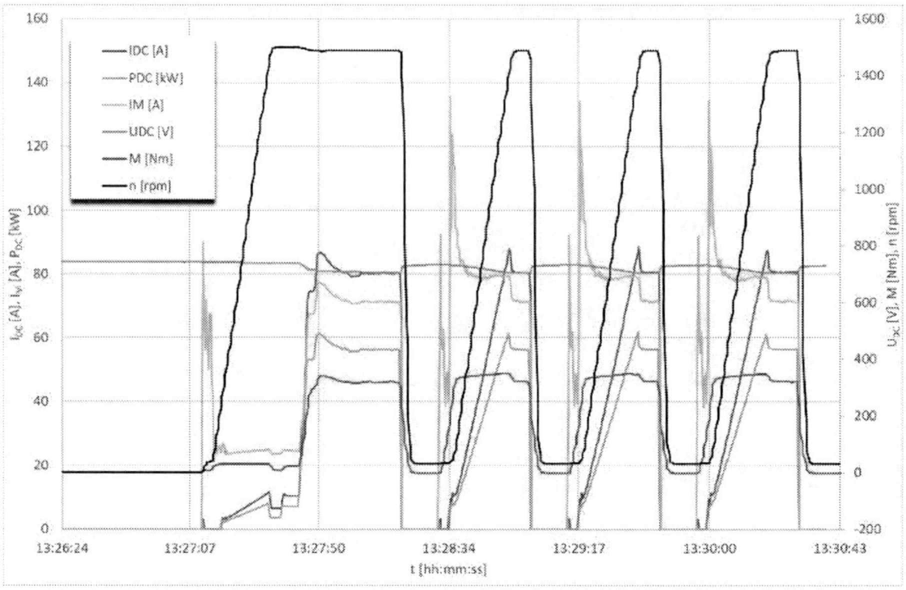

Figure 10. Exemplary time graph of mechanical and electrical parameters during the drive startup for 95% of the inverter rated current.

4. Summary and discussion

The tests of electrical properties of the floor-loader supply and control system with the autonomous battery supply, developed within the HYDKOM 75 project, are described in the article. The obtained test results confirmed the accepted design and functional assumptions. The control-supply system correctly realized charging of the cells with the current of 30 A intensity. The maximal charging time was 3 hours and ended automatically when an individual cell reached the maximal voltage of 3.65 V. In the drive mode, the system supplied the rated phase current of 75 A intensity, at the voltage of 500 V AC. The maximal time of the drive operation for the rated load was 1 hour and 6 minutes. After having reached the minimal voltage by a battery individual cell, the system of control and protections switched off the drive automatically, thus protecting the battery against the excessive discharging. An observation of the battery management system (BMS) forms the basis for meeting the requirements of PN-EN IEC 60079-0:2018-09 Standard in scope of using batteries in zones where explosion hazards occur [7]. Maximal temperatures of the selected system components, recorded during their operation in the charging mode and in the drive operation, do not cause any risk of gas or coal dust explosions in mine conditions. The startup properties of the applied drive enable a free motor startup for an initial load of 95% of inverter rated current.

References

[1] Kalita M 2011 Spągoładowarka górnicza jako wielofunkcyjna maszyna robocza z podwoziem gąsienicowym o szerokości 1000 mm *Nowoczesne metody eksploatacji węgla i skał zwięzłych TUR 2011*, pp. 270-278 (in Polish)

[2] Klich A 1999 *Maszyny i urządzenia dla inżynierii budownictwa podziemnego* Praca zbiorowa Katowice (in Polish)

[3] Przybyła H and Chmiela A 2002 *Technika i organizacja w robotach przygotowawczych* Wydawnictwo Politechniki Śląskiej, Gliwice (in Polish)

[4] Lesiak K 2019 Tests of flame-proof properties of supply-control system for a mining floor-loader *IOP Conf. Ser.: Mater. Sci. Eng.* 679 012013

IOP Conf. Series: Materials Science and Engineering 1134 (2021) 012005 doi:10.1088/1757-899X/1134/1/012005

[5] Kurpiel W 2020 Research on balancing BMS systems in a climatic chamber *Mining Machines* No. 3 (163)

[6] Deja P, Kapuściński D, Niedworok A and Polnik B 2019 *Badania innowacyjnego układu zasilająco – sterującego spągoładowarki górniczej* Maszyny Elektryczne - Zeszyty Problemowe Nr 2 (122), Sieć Badawcza ŁUKASIEWICZ Instytut Napędów i Maszyn Elektrycznych KOMEL, pp. 15-19 (in Polish)

[7] PN-EN IEC 60079-0:2018-09 Explosive atmospheres - Part 0: Equipment - General requirements

AUTHOR INDEX

Institute of Physics
Dirac House, Temple Back
Bristol BS1 6BE UK

ISSN: 1757-8981
ISBN 978-1-7138-3412-0